FORMAL THOUGHT AND THE SCIENCES OF MAN

BOSTON STUDIES IN THE PHILOSOPHY OF SCIENCE

EDITED BY ROBERT S. COHEN AND MARX W. WARTOFSKY

VOLUME 75

GILLES-GASTON GRANGER

Université de Provence, Aix-en-Provence, France

FORMAL THOUGHT AND THE SCIENCES OF MAN

With the author's *Postface* to the English edition (1982)

Introduction by Alexander Rosenberg

D. REIDEL PUBLISHING COMPANY

A MEMBER OF THE KLUWER ACADEMIC PUBLISHERS GROUP

DORDRECHT / BOSTON / LANCASTER

Library of Congress Cataloging in Publication Data

CIP

Granger, Gilles-Gaston.
Formal thought and the sciences of man.

(Boston studies in the philosophy of science ; v. 75)
Translation of: Pensée formelle et sciences de l'homme.
Bibliography: p.
Includes index.
1. Science—Philosophy. 2. Social sciences—Philosophy. 3. Languages—Philosophy. 4. Structuralism. I. Title. II. Series.
Q174.B67 vol. 75 [Q175] 501s [501] 83–2882
ISBN 90–277–1524–6

Published by D. Reidel Publishing Company,
P.O. Box 17, 3300 AA Dordrecht, Holland.

Sold and distributed in the U.S.A. and Canada
by Kluwer Academic Publishers
190 Old Derby Street, Hingham, MA 02043, U.S.A.

In all other countries, sold and distributed
by Kluwer Academic Publishers Group,
P.O. Box 322, 3300 AH Dordrecht, Holland.

This translation, by Alexander Rosenberg, of Granger's *Pensée formelle et sciences de l'homme* (Paris: Aubier, 1960; 2nd ed. 1967), was revised by Carolyn R. Fawcett and edited by Robert S. Cohen.

Printed in The Netherlands

TABLE OF CONTENTS

EDITORIAL NOTE

Gilles-Gaston Granger aptly named his workplace at Provence a 'seminar of comparative epistemology'. In this book, now a classic of post-war French philosophy of the sciences, Granger beautifully clarifies the puzzle of rational thinking about human beings. With a style at once rigorous and sympathetic, he calls upon all the resources of a philosopher and social thinker: Hegel, Husserl, Carnap, Piaget, Saussure, Herbert Simon, Lazarsfeld and Guttman, Homans, Mandelbrot, Kurt Lewin, Bergson, Marx, Freud, Bachelard. Granger's conceptual toughness is stimulating and also creative, and his critical reflections will bring the reader to understand what the author sees as central to social science: a 'dialectic of quality and axiomatization' (115). But we see further, in Granger's comment of 1982, that the matter is even more general, that a "conceptual preparation of the phenomenal field" must precede every sensible use of formal thought, that we must cope with the inevitable duality of rational thought, which is "the complementarity of an algorithmic manipulation and a conceptual determinism" (186). Only a dialectic, not any universal (which is to say, static) logic will do, will link the interdependent 'modalities of thought' which are logic and reason, symbol and experience.

Axiomatic reconstruction receives an impressive analysis at Granger's hand, with remarkable attention to the epistemological function and especially innovative power, even 'originality' (129), of this thoughtful process of intellectual labor which substitutes "a *simple idea* for a *common-sense idea*" (135). For Granger, axiomatization is instrumental and heuristic (136), an act that reveals the interrelations of hypotheses, and "calls forth ... a variation of models that is one of the essential conditions of the construction of effective and coherent concepts" (145). Whether Granger's detailed case studies of new and incomplete sciences, 'sciences *in vivo*' (176), persuade us that there can be a philosophically critical basis for the "difficult but fruitful marriage between combinatory and intuitive thought" remains open. This is so, even as to whether he has reached a sound schematic for how to think about the nature of the individual in his fascinating seventh chapter, with its meditative analyses of clinical judgment, information theory, and alienation, and his insightful *aperçu* into 'history as a clinical undertaking without practice' (168). But this is the character of philosophizing, and we are grateful for the rare fusion of modesty, maturity, and humane concern which this profound book articulates.

We are grateful also to Alexander Rosenberg for his introduction, and for his translation; and to Carolyn Fawcett for her impeccable work on all aspects of this volume, including the translation and the preparation of the bibliography and index; and to Professor Granger for his 'Postface 1982'.

ROBERT S. COHEN
MARX W. WARTOFSKY

Center for the Philosophy and History of Science
Boston University

January 1983

INTRODUCTION: GRANGER'S PHILOSOPHY OF SCIENCE

By and large, the Anglophone philosopher's acquaintance with French philosophy of science is limited to the reading of selections from the works of Henri Poincaré and Pierre Duhem. The importance that we attach to these philosophers is no greater than our ignorance of the works of their successors. One reason for this ignorance is, of course, that research in the philosophy of science, as English-speaking philosophers think of it, has never enjoyed the same prominence in France as it has among us; moreover, a great deal of what has gone by that name in Francophone philosophy has not addressed itself to those subjects, or employed those strategies that would attract our interest. Nevertheless, a tradition of philosophy of science, as we would call it, has persisted in France, and in recent years has produced a body of work which Anglophone philosophers can continue to ignore only to their own disadvantage. For this work offers a perspective on problems and issues with which we are concerned, but which is shifted away from our own just enough to make for new insights on these problems without being shifted so far as to transfigure the issues and insights into ones which we are neither concerned with, nor able to understand. It is these sorts of considerations which led to this translation of Gilles-Gaston Granger's *Pensée formelle et sciences de l'homme*.

Granger received his university education in mathematics and philosophy, and over the last twenty-five years has written, among other works, a major study of the conceptual and methodological situation of economic theory, *Méthodologie économique* (1955); an examination of Condorcet's social theory, *La mathématique sociale du Marquis de Condorcet* (1956); a general examination of issues in the philosophy of science, and especially social science, *Pensée formelle* (1960, second edition, 1967); a more advanced treatment of the role of mathematical and formal expression in science, *Essai d'une philosophie du style* (1968); as well as a translation of selections from the writings of Wittgenstein. He is professor of philosophy in the University of Provence, and has been a visiting professor at universities in both North and South America, most notably at the Institute for Advanced Study at Princeton. Over the years Granger has come to exert an important influence both within and beyond metropolitan France. Nevertheless, his views deserve a wider audience not only because of their intrinsic merit, but also because they represent the interests and the style of a French philosophy of science that is becoming increasingly relevant to our own concerns and methods. In order to foster this wider audience it seems advisable to offer a translation not of Granger's latest work at the very frontiers of his concerns in philosophy, but rather a translation of the introduction to his current view of science that the present work constitutes.

I

In general, Granger views science as a process, and the expression 'dialectic' looms large in his description of it. He writes, "It is . . . the scientific *movement* of thought which is the object of our study" (p. 2), not the system of a science which can no more be conceived as completed than the other activities of the contemporary mind. Indeed, science might well be characterized as "knowledge in error", for integral to science is the possibility of giving a precise meaning to an error, recognizing it as such, and basing further developments on this recognition. By way of examples, here Granger instances the nineteenth century's treatment of geometry, and in particular the axiom of parallels. It was the erroneous attempt to demonstrate this axiom that led to the development of the non-Euclidean geometries. Thus, "the progress of science consists, in part, of progressing from vulgar error – that is, from unformulated, ambiguous knowledge – to scientific error, that is, *to refutable knowledge* (p. 3)." This contrast between science and common-sensical "knowledge" expresses another of Granger's characteristic themes. On his view, the chief obstacle to science is entrenched common wisdom, especially insofar as its myths and prejudices are encapsulated in ordinary language. If we insist on describing phenomena in its terms, we will never rid ourselves of its burden of bogus commitments. There is in his treatment of ordinary language an echo of Bacon's critique of the idols of the marketplace, but Granger goes on to point out that it is the attempt to escape from the primitive conceptual net of ordinary experience that makes for the air of speculation and apparent practical irrelevance in so much of early scientific theory. It is when he turns to the social and behavioral sciences that Granger becomes most emphatic in his insistence that theoretical explanation must explicitly eschew the ordinary meanings and felt experiences that common sense appeals to in its explanation of human behavior. Instead, social scientists must find *structural* descriptions of the phenomena that they wish to explain. By 'structural description', Granger means the sort of categorical system reflected in Saussure's linguistic theory, and so influential in the great progress linguistic theory has made in this century. Indeed, Granger sees linguistic theory as expressing a paradigm for scientific theorizing, which research in other social sciences should adopt. But 'structuralism' as a method in science does not, in Granger's view, begin with Saussure and the linguists. It is nothing less than the strategy of all the sciences, both natural and social, since their beginnings. Now, 'structuralism' is a 'trendy' term no less in Anglophone methodology than in Francophone philosophy. But Granger's employment of the term is not to be assimilated to this trend, nor to the fashionable excesses for which this expression has been a watchword (he explicitly separates himself from this movement in the preface to the second edition).

The exact nature of what Granger calls 'structuralist' methods is the subject of a large part of this work, and I will not dwell on it much further in this introduction. Suffice it to say that Granger's demand for structuralist description is nothing less than the recognition that the successful pursuit of science requires that its terms and predicates pick out what we may call 'natural kinds'; that is, describe classes of items that bear uniform nomological relations to one another. A science whose descriptive terms do not meet this condition will never produce any laws that reflect such nomological connections. This means of course that the establishment of a vocabulary that reflects these natural kinds is not a necessary preliminary to the work of science, but is a constitutive feature of scientific progress, an activity that moves hand in hand with the formulation of the laws expressed in its terms. Moreover, since the vocabulary of ordinary language does not reflect these natural kinds, it follows that scientific theories must avoid its terms. Clearly, the natural sciences have succeeded in doing so. However, as Granger argues, the social and behavioral sciences have in many cases failed to overcome the obstacles erected in the ordinary descriptions of human behavior and have become ensnared in its scientifically sterile web.

At any rate, we may contrast Granger's basic views with both Continental phenomenologists who find the explanatory locus of behavior in experienced meaning and subjective interpretation, and Anglo-American followers of Wittgenstein who allege to find conceptual confusion in the search for behavioral categories beyond those of ordinary language. Thus, in terms of Anglophone philosophy of science, we can locate Granger's views squarely within the mainstream of post-positivism, but with some important additions. Just as he shows no trepidation about describing science as a 'dialectical' process, he is equally eager to employ another term uncharacteristic of

English-speaking philosophy of science: 'praxis'. By 'praxis' Granger means the characteristic technological activity and apparatus associated with and informed by scientific theory. In its controlling influence on the structure and goals of science, 'praxis' has, for Granger, much the same character as the logical empiricists and their successors attribute to experience. Granger, however, explicitly contrasts his notion of 'praxis' with that of perception, and claims that the former, not the latter, provides the content of scientific thought. The aim of science is, in part, the discrimination of objects, distinct from those which ordinary experience offers us, and which bear structural or nomological connections to one another. Whether a putative scientific object meets this requirement is largely determined through praxis, that is, through a human activity guided by a conception of this object: a technological, industrial, or practical activity. But theoretically informed technological activity offers more than a merely pragmatic confirmation of scientific theories. 'Praxis' not only confirms, but also fosters the invention and extension of scientific objects, in a way that merely passive observation could never do. It is the human need to control the environment, both natural and social, which leads to new technological arrangements of both men and machines. And it is these arrangements, so superior in determinateness and formal structure, that have enabled men to produce a natural science, and that increasingly foster social science. Granger offers some instructive examples of just how the establishment of technological relations between humans and machines may be exploited in the elaboration of scientific theory. This is a relation which Anglophone philosophers have hardly ever exploited or even noticed, assuming, as they often do, that the direction of causation is from theory to practice. Granger recognizes that like animal behavior, a human's action may be best studied under artificial conditions that restrict his responses. Modern technological society offers us the opportunity to pursue such studies, and Granger does not shrink from the suggestion that we do so. Here he betrays none of those concerns about 'dehumanization' that animate some critics of social and behavioral science.

II

Granger recognizes that "the study of the formal element in scientific knowledge cannot . . . proceed independently of an analysis of the function and structure of language" (pp. 6–7). Accordingly, he devotes the first several chapters of *Pensée formelle* to analysis of the employment of language in scientific theory. Although language has many functions, its principal scientific one is to convey information. And it is this fact, Granger claims, that vitiates much of the purely syntactic characterization of scientific theory. These characterizations are attributed to the early Carnap, and are rejected for failure to account for the informational function of language. Grammatical or syntactical well-formedness alone is no guide to the content of a sentence in any language rich enough for scientific purpose; whereas, contemporary information theory shows that a simple language employing no syntactical restrictions, and embodied only by a dictionary can communicate information, albeit inefficiently, simply through the concatenation of its semantic units. The introduction of syntactical restrictions reduces the range of possible concatenations of the semantic units, but thereby also diminishes the chances of misinterpretation of the informational content of the messages expressed in the language.

But the distinction between syntax and semantics is itself a dialectical one, shifting from level to level in the hierarchy of linguistic structure. Thus, consider written inscriptions. Inscriptions embody a set of graphic signs, which are governed by a system of rules that restrict their possible concatenations. The signs constitute a set of units that bear a semantic content, by contrast to the syntax that governs their relations. But each of the permissible concatenations itself reflects a distinct sound in oral language – in fact a phoneme; and in a sense these graphic concatenations express the semantic content of the phonemes, which are themselves 'contents' governed by yet another syntactic system that produces morphemes. These latter in turn provide the content for words, and words, of course, are governed by a grammar, strictly speaking. As Granger writes, "the *content* of the first level becomes the support of the *form* of the second; the semantic element becomes the syntactic element when the transition from writing to the spoken language occurs, a transmutation which is very exactly expressed by the transition from sound to phoneme. In my view, this is the general principle of the hierarchization of languages, which makes clearly apparent the relativity of the semantic and syntactic points of view in linguistic activity. All the more is this the case in the rigorous scientific execution of this activity" (p. 29).

Thus, Granger is interested in language not only because of its scientific employment, and because recent work in linguistics seems to him (as to other French philosophers) a discipline on the high road to scientific respectability, but also because the moving distinction between form and content at each successive level of language is in his view exactly mirrored by the interrelationship exhibited among successive scientific theories. It is his exploitation of analogies between science and language as described by structural linguistics that warrants the attribution to his views of the name 'structuralist'. Thus, Granger finds himself in that company of philosophers who, following Condillac, have described science as a language. But, unlike many of these philosophers, he recognizes that the semantic content of this language is paramount, not its logical form, and Granger is vigorous in his rejection of a purely formal analysis of science.

Granger finds one very important disanalogy between language and science. Ordinary linguistic activity is primarily vocal, and writing serves as a secondary code for transcribing this non-graphical activity. On the other hand, in science, written transcriptions are not mere codes, but the very substance of language. In science, the graphic language is primary, while the oral serves as an at best inadequate transcription. This is simply because graphic inscription offers literally far more dimensions along which linguistic representation can move than does the production of vocables. Through an account of the history of graphical expression in the science of chemistry, Granger documents the claim that it is by means of the successively greater exploitations of these spatial dimensions of expression that scientific theories find their canonical form. Even more strikingly, the development of graphical symbolism facilitates new factual discoveries, as well as suggests the direction of further research: it is "in the conquest of *dimensional pluralism*, permitting the simple and fruitful expression of complex structures by means of which science explains experience" (pp. 37–38). The fundamentally graphical character of scientific language also fosters the treatment of scientific theories as abstract uninterpreted calculi, and thereby provides them with what Granger calls 'semantic polyvalence'. Because we can often provide alternative domains of objects that satisfy these abstract calculi, and thus extend the syntactic range of a theory's language, syntax comes increasingly to be the most significant element in the language of science. But, though this fact provides the motivation for the all-too-formal treatments attributed to Positivism, it must be remembered that the function of scientific as well as ordinary language is the communication of information, and for this it must have objective reference. This reference, which is tantamount to an interpretation or semantics for this syntax, is provided in scientific praxis, in the *découpage* of phenomena.

III

'*Découpage*' is one of the few words that has been left untranslated in the present work, for it expresses an especially central notion for Granger's philosophy of science, and there is no single English term that captures its precise meaning and flavor. *Découpage* has an English cognate, and this cognate sheds some light on its meaning in Granger's work. *Découpage* is the technique of decorating a surface with cut-outs, and 'cut-out' is the most direct and the simplest literal translation of the French '*découpage*'. But by this term Granger means to suggest that out of the amorphous material of experience, science must separate, mold, shape, contour, arrange and refine its objects. This *découpage* must be an active process, involving the coordination of theoretical and practical results in the delineation of objects which can satisfy the syntactical restrictions characteristic of scientific language. In fact, as noted previously, the activity of *découpage* is nothing less than the determination of the natural kinds which scientific theory treats. Granger does not employ the expression 'natural kinds'; rather he talks of 'the scientific object'. And the aim of the scientific *découpage* is the substitution of scientific objects for the objects 'given' in ordinary experience, and our everyday conceptual scheme. Turning to behavioral science, he argues that the history of linguistics in this century has been the history of just such a succession of *découpages*, which, informed by practical needs and technological demands, have resulted in the construction of at least one set of undeniable scientific objects in the human disciplines. The key to the construction of a scientific object in linguistics was the attempt to find a characterization of types of sounds, not in terms of the common physical or semantic/syntactic properties of their tokens, but in terms of the interrelations of their phonemic functions. These interrelations, first hit upon as significant by Saussure, constitute a structure, which is the object of linguistic theory, and which through the years has come to underlie well-confirmed descriptions of both cross-sectional and developmental phonologies. Granger details this history, and then connects it with technological developments that both confirm and extend this work in a way that clearly manifests Granger's insistence on the importance of scientific praxis. In particular, he cites work in machine translation, and in the production of systems which can convert speech into visual data that can be 'read' by the deaf at a rate comparable to normal hearing and with accuracy no worse than a poor telephone connection.

Work in structural linguistics Granger describes as formalistic *découpage*; even though it comes soon enough to have practical connections, its origins lie in a purely theoretical context. By contrast, the 'operational' *découpage* begins with problems of applied science and moves ultimately to formal, structural theory. Granger's example is *Operations Research*, and he argues that at least three of its main lines of research reflect a movement from limited practical problems to the same sort of structural theories that the development of linguistics has exemplifed. Thus, he finds in areas like queuing theory, game theory, and linear programming, the elaboration of a structurally characterized scientific object: the notion of *decision*. This concept must, however, be understood in a sense quite different from its ordinary one, if it is to have a systematic role in the sciences of man. For, as Granger concludes,

> ... the scientific revolution in the domain of mankind consists first of all in freeing itself from the naive modes of *dècoupage* transmitted by ordinary language. Human events, taken at the level of experienced meanings, can give rise only to a pseudo-science, a more or less skillful discourse which only reflects an empirical practice, even if it is raised to the level of an art. The transmutation of the phenomenon into an object is achieved by the convergence of two movements which cause forms to penetrate into the world of events ... [the] formalist *découpage* ... aims directly at the construction of abstract systems which it studies apparently for themselves. [Whereas in] Operations Research ... formalization is subordinated to a perspective of action. But ... the two movements presuppose each other and rejoin, offering a glimpse of the possibility of an original discipline ... (p. 84).

IV

Granger recognizes that the chief obstacle to a scientific treatment of human behavior and action is the allegedly qualitative character of the phenomena with which such a science must deal. Opponents of the extension of the methods of natural science to the social studies argue that the subject matter of the latter disciplines is not open to the sort of quantification characteristic of formal thought. According to Granger, this sort of argument rests on at least two mistakes. The first and most obvious is the supposition that the formal thought characteristic of natural science is nothing more than the manipulation of metrically quantifiable variables. The view that nothing is susceptible of formal or mathematical treatment unless it is open to direct or indirect measurement involves a completely artificial restriction on the range of mathematical expression and a good deal of ignorance of the non-metrical foundations of contemporary mathematical thought. By way of an example, Granger summarizes the treatment of linear structures and vector calculus advanced by the Bourbaki group; he concludes that metrical "quantities figure in [this formalism] only as auxiliaries, and naive thinking is tempted to identify the elements of vectorial space with magnitudes, by implicitly endowing them with predicates tacitly recognized by [their] intuitive quantities" (p. 111).

This tacit appeal to the intuitive is, of course, the second of the mistakes that the objection from quality rests on. The erroneous supposition here is that to establish the *possibility* of a science of man requires nothing less than the actual provision of a theory which meets scientific standards, and accounts for human behavior by reference to the very qualitative features or contents of our experience that ordinary common-sensical language and thought appeal to. Granger recognizes that the contents of experience, the phenomenological richness and symbolic meaning with which everyday life is imbued, cannot be captured by science: "A science of man can naturally replace neither the fine arts, nor the concrete individual practice of human relations. No more than physics, or chemistry can be substituted for the 'flavor' of sounds, smells, colors, or for the art of cooking" (p. 116). True, it cannot do these things, but in order to accomplish its task it need not do them. Formal thought need not reproduce quality, rather it must attempt to regiment otherwise isolated qualitative differences into coherent structures, into patterns of systematic contrast and opposition of the sort which characterizes linguistic reduction of phonemes. Like such a reduction, this structuring does not deprive its objects of their qualitative properties, but it refrains from classifying and organizing them by reference to these 'internal' and 'isolated' features. And this explains both why the application of formal thought to human action does not require the measurability of qualitative content of its objects, and why metrical and non-metrical mathematical techniques available from physical science are suitable to the description of structures in social science. Granger shows how results in psychological scaling and factor analysis substantiate this claim in an examination of the work of Guttman, Lazarsfeld, and Stouffer. The aim of these social scientists is to find a framework within which qualitative data (such as might be made available in a survey research study for example) can be organized so as to reveal systematic interrelations that can be formally expressed in terms of testable hypotheses. Their techniques do not ignore the qualitative data, but provide an initial and partial organization of it, by reference to its broadest categorical features. This first structuring then provides the basis for successively finer categorization, and unification, which at each level progressively 'lessens' the degree to which qualitative features remain isolated and unconnected. The succession of structures that account for the qualitative aspect of human phenomena expresses for Granger the dialectic movement to be found in natural science as well, and mirrors the syntactic/semantic shifts in the hierarchy of language.

The reduction of the qualitative element of human experience to structure essentially involves the sort of thinking typified in the axiomatic approach to the foundations of mathematics, and Granger attributes to axiomatization a very special role in the methodology of social science. After examining a variety of functions which axiomatization plays in mathematics and natural science, he argues that, to a much greater extent, axiomatization is an instrument of discovery and confirmation in the sciences of man: "The functions [of axiomatization] that we have noted in the domain of physics – destruction of pseudo-evidence and [the provision of] experimental articulation – here [in social science] converge to rectify embryonic scientific thought, which is too easily blinded and confused by the brightness and glitter of experienced meanings. These attempts at axiomatization, however awkward and partial they may be, awaken thought from its repose in common sense. They offer themselves as explicit experiments of *eidetic variations*, carried out on initially shapeless notions, from which the minimal conditions of coherence and efficacy are extracted" (p. 137).

While axiomatic presentations are characteristic of 'completed' theories in the physical sciences, and represent a point of *culmination* in the elaboration of such theories, in the social and behavioral sciences attempts to axiomatize are necessary *initial* steps in the development of a theory. For axiomatization offers "rational thought the sole means of escaping from the attractions of data derived from experience" (p. 145). The propensity to find scientific explanation in the appeal to the phenomenologically or vernacularly given is much weaker and easier to resist in natural science or mathematics. These disciplines turn away from ordinary language and experience, construct their objects, and *then* reach the stage where axiomatization is important. But where the inclination to appeal to common notions is stronger, the process of axiomatization is a first step which provides a "drastic and certainly aggressive asepsis" (p. 146) of these notions. The eventual aim in social science is the integration and interanimation of local axiomatizations of behavior into global axiomatizations akin to those of natural science. Although Granger recognizes that it is far too early to find such interarticulation, he does offer an examination of developments moving in this direction.

V

It is in the last chapter of *Formal Thought and The Sciences of Man* that Granger turns to matters that are largely foreign ground to Anglophone philosophers. For in this chapter he takes up the question of how science, which is unavoidably general in its expression, can come into contact with the individual human agent. He rejects history as a *science* of individual human action, for history proceeds by the use of those very terms of ordinary language that social science must endeavor to ignore. Its goal is not the constitution of a new scientific object, but the reconstruction of the past in terms of a contemporary ideology. It is in the clinical practice of medicine, and in particular in the practice of the psychiatrist that Granger finds the avenue of approach for science to the individual. But in setting out this view Granger is at pains to show how far contemporary psychiatric (particularly psychoanalytic) practice and pretensions fall short of the requirements of science. He appeals instead to a structural account of personality theory which exploits analogies with successful theories such as those of linguistics.

With this treatment of a science of the individual, Granger brings his study to a close, recognizing the tentative character of both contemporary social sciences, and his own account of its progress. How tentative either or both in fact are, is a matter that can only be left to the future.

Syracuse University ALEXANDER ROSENBERG

PREFACE

TO THE READER: ON STRUCTURALISMS (1967)

This book was first published at a time (1960) when the words structure and structuralism did not yet enjoy the universal repute they now have with writers of journals and newspapers. I say: words, not realities; and I dare hope that this work has in no way contributed to the diffusion and confusion of meaning which, in such cases, accompanies the fortune of the words. The purpose of this preface is to take, briefly, a stand on this point.

It would be good to begin by underlining the importance of the original plurality of the forms of structuralism. Besides, to speak of *the one* structuralism hardly makes sense, not only because of the tendency today to enlist every fashionable writer under that flag, but also because the term itself, even when legitimate, corresponds to quite significantly distinct orientations. *The* structuralisms, then, *in the plural*, derive from three different points of the contemporary universe of culture, each one intended to designate a certain fundamental object of knowledge with the name 'structure', each one opposed in its way to the previously adopted paradigms of knowledge. The histories of philosophy, mathematics and linguistics have been the three foci. To say that these three aspects of structuralism have remained distinct in no way means that we are denying the existence among them of certain relations of influence. But the difference of fields and people means that the monolithic vocabulary tends to conceal here a multiplicity of particularly instructive viewpoints.

Everyone knows that the conceptions of the linguistic object expounded first of all by Baudouin de Courthenay and de Saussure are at the source of the structuralism of linguists. Its governing idea is that language, considered independently of the entire context of concrete activities of expression and their historical evolution, constitutes a legitimately *découpé* object of science, forming a *system* whose intrinsic determinations can be described as such. Once the reduction of the phenomenon to the abstract object that is *language* has been effected, the second Saussurian idea assigns this object its nature; perhaps this could be expressed roughly by saying that each of the elements of the system of language can be defined only in terms of its relations of opposition to all the others; each of them is somehow 'uncolored'[1] and assumes value, function and meaning only relative to that from which it is demarcated within the entire system. All this is well known today; it is still a good idea to state that all contemporary linguistics, even when it refuses allegiance to Saussurian structuralism, is ineluctably dependent on it in that only the idea of a language as system makes it possible and gives it its basis. More generally it seems to me that the basic idea of Saussurian linguistics is still that it gives unity and meaning to every structural doctrine. A strong, simple idea according to which every attempt to know anything of man objectively must first pass through a reduction of experience to a system of correlative marks.

But however original and fruitful the Saussurian idea has been, it could not be presented as constituting in itself alone the core of the structuralisms. Completely independently and in a universe of thought alien to that of Saussure's first disciples, a notion of structure was delineated in the thirties through the effort of the Bourbaki mathematicians to give an adequate description of the form of their science as it has taken shape after the end of the last century. The idea that is essential, and at its foundation common to the mathematicians and Saussure, is that the object is perceptible in its depth not so much as the bearer of inner properties – in the image of perceived qualities – but as the system of relations between elements not otherwise marked, whose only envisaged properties derive from these relations themselves. So that the true object of mathematical knowledge is the structure, not the element: what the analyst aims at, for example, when he states the properties of complex numbers, are the formal properties of a *system* of objects which he sums up under the name of the structure of an algebraically closed commutative field. Each branch of mathematics thus explores a structure, or a complex of structures; and the very notion of structure in general can receive a rigorous definition,[2] assuming as given, however, the instrument of naive set theory, and the mathematical usage of the series of integers. Without entering in any way into a discussion of the definitive or provisional character of the now classical construction of mathematical structures, it seems that in this field the essential structuralist idea, namely that knowledge of a mathematical object consists not in the isolated qualities of an entity but in the formal properties of a system, can be considered an accepted fact. And the nature of the relations that each time determine the system is such that they can be described and recognized unequivocally. It would no doubt be a good idea to reserve the name 'structures' for such systems; but if one does not succeed in doing this – and I myself would certainly not claim to succeed – one must at least always stress the possible consequences of this homonymity and the specious analogies it tends to foster.

It is precisely the existence of a third source of structuralism that compels one to demand vigilance on this point, not because structuralism in the history of philosophy appears to me a distortion of meaning but, on the contrary, because it obliges one to dissociate the structuralist intention of the objective constitution of structures. When Martial Gueroult published his *Descartes selon l'ordre des raisons* in 1953, he proposed, independently of linguists and mathematicians, the first monumental example of structural analysis of a philosophical work. What does structural mean here? It seems to me that the method, if not the art with which it is used here, can be defined quite simply. The structuralist idea in the history of philosophy consists in considering a work in itself, as a relatively closed and autonomous system which the analyst wants to understand as such. Thus the Saussurian idea of language is rediscovered and applied to a phenomenon of culture at once less extensive and more complex – since it assumes the former. Of course, it is permitted to transpose this idea to other works, seeing that it is possible to postulate their systematicity without improbability. But what is the nature of the system here? This question is answered by the second regulating principle implicit in this structuralism. It should be noted, parenthetically, that I am not here interpreting the acknowledged intentions of Martial Guéroult (or Victor Goldschmidt or Ginette Dreyfus or Jules Vuillemin in *Physique et mètaphysique kantiennes*), but the historian's own approaches. The philosophical system that he outlines is a set of certain logically connected thoughts, but in truth not at all assimilable to the rigorously abstract and closed[3] system of a mathematical structure or a phonological structure in the field of the linguists. The elements of the system here – described at the level of propositions or at the level of concepts – are open and always incompletely determined by their reciprocal relations. Hence it happens that every rigorous attempt at the axiomatization of a philosophical work leads precisely to giving rise to the impossibility of an integral formalization. I would say, then, in my own language, that the second principle of structuralism in the history of philosophy consists in positing that the elements of the system are 'significations', not 'meanings'.[4]

These are the three sources of contemporary structuralisms. But what a broad public designates by this term is sometimes connected to it only by the untimely use of the word 'structure', or even by abusive references to a superficially examined linguistics or a misunderstood mathematics. My purpose here is not, however, to polemicize or to cry, at my risk, that too often *the emperor has no clothes*. I only wanted to propose the quite reasonable hypothesis of a genealogy of the structuralist idea, a genealogy that brings to light, on the one hand, the fundamental theme that justifies its unity, and on the other, the diversity that is the source of many a paralogism.[5]

Perhaps we now see more clearly how to try to distinguish the positive traits of a structuralism through the epistemological problems it poses, beyond the pleasures of modernity.

Since we are dealing here, of course, with the application of the structuralist idea to the human sciences, a first point can be briefly called to mind, questioning the preponderant place often accorded structural linguistics as a scientific paradigm. We just recalled its decisive importance in the formation of the idea; does it follow from this that it is necessary to consider every structuralist conception of knowledge as modeling knowledge on linguistics? It seems to me that the answer must be negative. Models other than the phonological system seem concurrently usable and reasonably adequate. I shall not develop this thesis here, which, prior to the drafting of the present work, I have tried and shall try to sharpen elsewhere. But if 'panglottism' is a childhood disease of structuralist thought, it is nonetheless a good idea to recognize that it only pushes to extremes a correct idea, namely that on the one hand all science is necessarily produced within a language, and that, on the other, all human work is presented, at least at one of its levels, as a signifying system. The first clause means that there is no science without articulated symbolism; the second, that every science of the human fact must recognize in it, beyond the various possible 'energetic' organizations that constitute it as a functional machine, one or more symbolic organizations.

It is here, moreover, that the essential epistemological problem of structuralism arises: what is the nature of the organizations it postulates as constituting the object of a body of knowledge? The three sources that have been indicated serve to give an account of the confusions that can take place. To give the very name 'structure', incautiously, to logico-mathematical systems, to organizations of the phonological type, and to the conceptual texture of a philosophical discourse, is to open up a possible path to nonsense. It is quite true, however, that each of them presents a paradigm admissible of systematicity. But it is up to a philosophy of knowledge to describe them and distinguish them, whether one believes that one can conclude in favor of the equal validity of each for the constitution of an object, or whether one succeeds in showing the illusory character of certain of the three meanings. The danger of the term 'structuralism' lies in the fact that it leads one to understand that the problem is already resolved in the sense of univocality.

Now, once the confusion is made, it makes impossible the elucidation of the most general question posed by the worldly success of structuralisms: does it deal with a vision of the world? Under the original forms that have been recorded above, the answer is not in doubt: structuralism is a methodology, or more exactly, it is presented as a bias toward the determination of the constitutive categories of an object of science, and particularly of the object of the human sciences. But this formulation already shows that the structuralist position involves more than a simple choice of method, since it postulates a certain definition of what is objectifiable in man and knowable by science. It is then quite natural to associate with it the position of theses relative to the nature of reality. This is an association that, for my part, I nonetheless persist in finding mistaken, despite the guarantee that Claude Lévi-Strauss, up to a certain point, seems to give it; Lévi-Strauss's epistemological positions are apparently completed by a thesis on the nature of the mind. But what counts in fact is the meaning of systematicity that one chooses in order to define the object of knowledge and this alone depends on the structuralist idea. The philosophical meaning of such a position would consist first of all in the fact that every objectification of experience is defined as structure – which eliminates all philosophy of knowledge of the 'intuitive', affective or 'mystical' type; these types, although by no means denied as aspects of experience, become, as modes of knowledge, examples of non-knowledge. Following this it would remain for us to decide – and this decision could not be positive in nature – whether the structuration that objectifies can depend on several meanings of systematicity. Thus one can conceive of a 'mathematical' structuralism that is objectivization only by reduction to structures in the strict sense, as we have brought out above.

Such a doctrine is hardly anything but a derisory scarecrow set up by the adversaries of science to deflect people from a positive knowledge of man. And even if one would find a genuine declared representative, it would not be good to confuse this 'mathematism' with an ontology: to postulate that every *object* is such only by means of a reduction to *mathematica* does not mean directly that *mathematica* are beings-in-themselves and primary. If, on the contrary, one admits a pluralism of modes of objectification, recognizing as legitimate, for example, the constitution of objects of the 'semantic' type (in the sense of phonological systems which would be their prototypes),[6] namely 'objects' of the type of 'systems of significations', it is good to provide the means for never confusing these different concepts of systems. To treat, for example, a structure of significations – such as a philosophical system, the organization of a psychism, the ideological substructures of a culture – as a mathematical structure,[7] by pretending to attribute identical properties to them, when there is need, would be to fall into the transcendental illusion where those who, as Heraclitus said, *shoot arrows into the darkness* are indeed happy. But this critical distinction in no way leads to otherwise denying the possibility of a superposition of structures of different orders for the reduction of the same phenomenon: thus in the second case cited nothing hinders one *a priori* from envisaging a psychic organization on the one hand as a system of significations, on the other hand as a 'semantic' system, and finally as a mathematical structure.

As the design of the present volume has above all been critical, it was not a matter of showing that a structuration in the strict sense was possible and fruitful for the object of the human sciences. And I still think today that only the last two meanings of structure, as they have just been indicated above, are susceptible of constituting true *objects*; as for the first, the systematicity of significations does not seem to me to really deserve the name of 'structure'; it would then be constitutive not of the object but of what makes the theme of philosophical interpretation. This is why, in any case, the discourse of the book and even the book itself in no way claim either to depend on a scientific knowledge or to determine an object. They explore or try to explore the *significations* of a certain scientific object and to outline its system.

Without a doubt I would write this book today in another way, or more exactly, an altogether different book should be written. Nevertheless the same orientation would reappear whose formula I wanted to make explicit to the reader in this preface, as it looks to me today. What is more, what I have tried and shall try to publish afterwards will make little real sense except on the basis of this work. Hence I have preferred – but perhaps it is also laziness – to let it be reprinted as is, reserving the modifications and extensions I want for new writings.

GILLES-GASTON GRANGER
Jouques, 15 October 1967

CHAPTER I

THE PROBLEM OF FORMS AND THE PHILOSOPHY OF THE SCIENCES

1.1. Let me justify the title of this essay right from the start by sketching out what appears to me to be the current fundamental problem for a philosophy of the sciences.

In order to understand the sense and trends of the philosophy of the sciences, a complete genealogy is hardly necessary. Nevertheless, one cannot avoid taking one's bearings from a reference point in the philosophical past: Kantian criticism, which remains a sort of center of epistemological diffusion. But it is important to be clear about the uses to which history will be put here. I am not a historian, but a reader, a 'consumer', with respect to the philosophies of the past. Naturally, then, I shall run the risk of misunderstanding them, since I read their works with my own preoccupations, and as if they really should instruct me. The patient analysis of texts, which is the historian's enterprise, is my guide, and not my goal.

Accordingly, the brief considerations which follow should in no way be taken as the work of a historian. The historian's task is to understand, rather than to interpret, as Vuillemin put it so well in his work on Kant's physics. Thus, when a philosopher's themes and methods of thought are invoked, it will not be in order to better comprehend them, but in order to test their force and their implications in connection with the current features of a problem. Although this attitude is utterly unsupportable by the true historian, for us it can be legitimate and fruitful, if, through this testing, it allows us to better delineate the nature of our difficulties, to define more completely our real problems, and to question our prejudices. Clearly, it is not in order to produce an eclectic construction (bound in any case to be disappointing) that one undertakes to compare disparate philosophical methods. In the perspective of a philosophy of the sciences, what the historian offers us when he explicates Descartes or Kant is an instrument of *dissociation*, rather than of fusion – or of confusion – of concepts. Historical elucidation enables us to understand the method of thought of philosophers; if a modern interpretation of the problems of science makes use of these instruments – as an attempt to use ancient tools on new material – it is in order to better reveal the contours of these current problems. Such an application of the philosophical past may not be a valid way to proceed if our desire is to enrich our knowledge of history, since in this regard it risks complete confusion and travesty. However, we cannot be accused of falsifying historical truth, for we intend to apply the methods of a philosopher to data inaccessible to his time only occasionally, not in order to extend his doctrine, but rather to better clarify our own obscurities. Perhaps there is no other means of measuring progress in philosophy than this possibility of always confronting an old mode of thought with new data, a confrontation which cannot fail to show both how instructive the old philosophy remains, and yet how circumscribed it appears in the face of an everchanging universe.

1.2. Turning to the interpretation of science, Kant, in my view, introduced the original and most durable and fruitful manner of posing the problem. He oriented modern epistemology, if not in its content, then at least in its form, by questioning the possibility of science. Today this is still the question to be posed. But we should not demand solutions of the critical enterprise. It suffices that the critical enterprise bring to light the themes which the current state of science invites us to meditate upon.

THE *POSSIBILITY* OF SCIENCE AND THE *FACT* OF SCIENCE

1.3. If the problem is to state how science is possible, then there is a great temptation to consider scientific knowledge as a norm, an idealized reflection of one of its stages. The philosophy of the sciences would then be the hermeneutics of a mythology. Certainly, it is true that science assumes this existential character of myth in our consciousness and in our mores; but thus considered it depends on a sociology and a psychology of knowledge. The epistemological attitude does not concern itself with this reflection of science in the individual consciousness or in social life; rather it aims at the *practice* of science, in its process of creation, and its employment. Naturally, this practice involves at each stage an ideal of knowledge, but it is important not to confuse this ideal, which is an integral part of scientific thought as *fact*, with a universal and predetermined norm. Science exists in fact; the most paradoxical preliminary difficulty of epistemology is to grasp science as such, without substituting for it an hypostatized image. It is thus the scientific *movement* of thought which is the object of our study, not the system of a science implicitly considered complete, nor the particular works of the contemporary mind. Doubtless it is not easy to refuse this double temptation. Nevertheless, science exists. It does not reduce to an ensemble of definitively established, rationally interconnected dogmas, as the logic of Aristotle appeared to Kant, "to all appearances finished and completed." Although the criterion of authenticity of the scientific spirit appeared to the philosopher of the *Critique* as the attainment of definitive systematization, today it does not seem necessary to us to require this mark in order to recognize the "sure route of science." The scientific edifice is necessarily in disequilibrium and always in progress. In this progress error does not merely play the role of psychological accident. It is an integral part of the movement of mind that gives rise to science to such an extent that one might conceive of defining scientific knowledge paradoxically as *knowledge in error*. By this we mean that science involves the possibility of giving a precise meaning to error, recognizing it, and basing new developments upon it. Thus, the mathematicians of the seventeenth century believed that they could demonstrate the fifth postulate of Euclid; indeed, they offered demonstrations of it, but in error. But this ill-founded pretension expressed itself and developed in such a way that it became possible to expose its exact presuppositions, and this analysis of knowledge in error was the point of departure for a revision of the science of geometry. One could easily find well-known examples of this process in history, and there is no need to insist upon this trivial fact. But perhaps it has not been sufficiently noted that there is in science a special maturation of errors. One might even say that the progress of science consists, in part, of progressing from vulgar error – that is, from unformulated, ambiguous knowledge – to scientific error, that is, to *refutable knowledge*. In this sense, science continually destroys itself in order to be reborn, or perhaps more correctly, to be truly born.

1.4. In consequence, the search for the conditions of the possibility of science cannot consist in an *a priori* description of closed transcendental forms that outline the blueprint of all scientific knowledge. It is from the perspective of this rejection that I should like to adopt the statement of Jean Cavaillès, that epistemology requires not so much a philosophy of consciousness as a philosophy of the concept. This philosophy of the concept would be nothing other than the interpretation of the progressive embodiments of scientific 'error'. It is not to be confused with an anecdotal history of theories, for scientific progress is not to be identified at all with the conceptual fluctuations that still depend on the psychology of scientists and the sociology of scientific knowledge. More than science itself, it is scientific ideologies, i.e. the reflection of science in the consciousness of a group or class, which depends on the these factors. Whatever may be the importance of these ideologies, we believe that it is nevertheless permissible to take science in itself, and epistemological reflection can be justified only if the systems of scientific thought reveal an order of reasons, which, without conferring on them any absolute autonomy, nevertheless manifest the authenticity of the movement from which they proceed.

Science, as we take it as a theme of reflection, is thus science stained with errors and inadequacies, it is science *de facto*, and not science *de jure*, [which is] imaginary. And it is not merely in its content that science gives an only illusory definiteness: of course, no one would ever dream of denying this, but it is illusory in form as well. Should all attempts, then, to provide a transcendental determination of the scientific object be rejected as impossible? Yes, if by transcendental one means a definitive system of the conditions of objective knowledge. No, if one recognizes that science, although necessarily defined by *a priori* conditions which it gives itself, does not give itself these conditions under the form of closed systems, but constantly varies its requirements. Thus, one can speak paradoxically of a *transcendental progress*, which is philosophically – and even technically – more significant perhaps than the progress of the contents [of science], from which it is in fact inseparable.

PERCEPTION AND SCIENCE

1.5. It is plain that, for Kant, the conception of a definitive transcendental determination of the scientific object derives from the identification which he postulates between the perceived phenomenon and the object conceived by science. The whole enterprise of the *Critique* presupposes a radical homogeneity between the forms of perception and the forms of scientific knowledge. As one believes oneself justified in acknowledging an immutable transcendental system of perceptual activity, so one describes an object of science as determined *a priori* by the principles of the grasp of experience. But the transcendental philosophy of knowledge is not really involved in the projection of the results of an analysis of perception onto the activity of scientific conceptualization. Rather, it seems that the determining step here consists in taking mathematics as a work of pure sensation, and in drawing out of geometry the monogram of all objective perception. The result is that, on the one hand, the apprehension of sensations is through the forms of Euclidean geometry, taken as the prototype of all mathematical activity; on the other hand geometry finds itself prisoner of sensible intuition. It was in this narrow and ambiguous sense that Kant was able to write " . . .

in every special doctrine of nature only so much science proper can be found as there is mathematics in it" (Kant 1970, p. 6). This was because for Kant perception itself is an immanent mathematics. To be sure meditation on this statement of Kant's has never ceased, and the work that we are pursuing here is in a certain sense nothing but a development of this theme. Yet we are immediately faced with the fundamental problem of making precise the relations between mathematics and perception.

1.6. If it is true that the object is scientific only to the extent that it depends on mathematics, it is not because mathematical thought is the simple systematization of the forms of sensory perception. Quite the contrary, the transcendental attitude of analysis leads us to recognize that mathematics draws us further and further away from the thing perceived. There is no question of retracing here the psychological origin of the structures of arithmetic and geometry, such as has been examined with great interest in the fine work of Piaget. We are considering science at the level of adult thought and observe that every object which appears in the discourse of science presupposes a mathematics, either implicit or explicit, but one whose ties to perception seem to be weak. The transcendental attitude shows us the strategic role of mathematics in knowledge, and it does so with such vigor that we are brought to define all scientific forms of thought as mathematical. But at the same time it confronts us with a new difficulty: for it is no longer possible to conceive of mathematical forms as simple thematizations of schemes immanent in the perception of the sensible. What then are these structures, what do they signify in relation [to the *sensible*], to the thing perceived, what is the nature of the objectivity which they constitute? The problem for a philosophy of the sciences takes shape, then, while the original transcendental project is shifted; the spontaneity of objectivity in perception is contrasted with the laborious search for scientific objectivity. Science is human labor.

LINGUISTIC EXPRESSION AND SCIENTIFIC FORMS

1.7. The problem of an epistemology is thus a problem of the interpretation of forms. It is just the same for a theory of perception, as much in its experimental aspect as in its philosophical one. Why is it then that one is not at all the extension of the other, and that there must be a leap from the object of perception to that of science? Because the intervention of form in the object of perception is spontaneous and immediate, while in the object of science it is a laborious and relatively contingent mediation, like a work of art. The form of the scientific object does not directly involve sensible content, but a *language*. We cannot point out enough how much the consideration of linguistic formulation is passed over in utter silence in the Kantian corpus. The mediation of language, which plays such an important role in Leibniz's philosophy of logic and in the empiricism of Hume as well, has no assignable place in the orthodox critical perspective. Moreover, once its legitimate role is made clear, the postulate of the homogeneity of perceptual form and scientific form can no longer be maintained: the hiatus between perception and science is due essentially to this mediation of language. It might thus be said that the most significant contribution to an epistemology based on the Kantian approach is that of the neopositivists, with the extraordinary developments of their linguistic analysis of knowledge.

1.8. But the neopositivist discovery led immediately to a nominalism of which the most extreme formulation is certainly Wittgenstein's. Now, this nominalism condemns us to understand nothing about the true progress of knowledge. How can science be reduced to a language without denying it any power over things? We refuse to content ourselves with the artifices of nominalism in order somehow to join purely linguistic understanding to an effective grasp of nature. Better yet are the enigmatic lyricism and metaphysical humor of Wittgenstein, for at least they show up the difficulty:

> My propositions serve as elucidations in the following way: anyone who understands me eventually recognizes them as nonsensical, when he has used them – as steps – to climb beyond them. (He must, so to speak, throw away the ladder after he has climbed up it.)
>
> He must transcend these propositions, and then he will see the world aright.
>
> What we cannot speak about we must pass over in silence [Wovon man nicht sprechen kann, darüber muss man schweigen.] (Wittgenstein 1974, *6.54, 7*).

But no. It is precisely because at every moment man encounters that which he has no immediate means of speaking about that he improvises a science, invents a new language, and annexes new objects.

In any case, the problem is posed: science apprehends objects by constructing systems of forms *in a language*, and directly on the sensible given. How, then, is the effectiveness of its impact in the perceived world guaranteed? This is the essential theme of epistemology.

1.9. The study of the formal element in scientific knowledge cannot thus proceed independently of an analysis of the function and structure of language. It is dangerous, however, to consider linguistic constructions independently of the objective aims to which they are naturally ordered in the work of science. The nominalist and the formalist temptations are assuredly the most dangerous shoals of a philosophy of knowledge, which must take account of language. The exclusive concern with an analysis of linguistic constructions leads one to mistake shadow for substance, and to substitute grammatical constructions for the objects of science; a formal epistemology which is not at the same time transcendental, which makes no attempt to describe the modes of determination of real objects, cannot satisfy our desire to understand science, and nature.

1.10. No doubt it is Husserl who deserves the credit of having directed epistemology along the difficult path of research at two levels: that of language and that of objects. To approach the problem of science it is essential to recognize that the horizon of the scientific *logos* is radically different from the horizon of perceived objects. Neither words nor phenomena are in fact actual and strictly delimited entities. But the nature of their implicit contents is so profoundly distinct that any interpretation of science as a homogeneous extension of perception will lead us utterly astray. Thus, on the one hand it is necessary to try to investigate this horizon of language in its scientific usage; but on the other hand the articulation of language and the immediate perception of things must also be elucidated. The merit of phenomenology is that it has clearly expressed this two-fold task. From the point of view of a philosophy of the sciences – moreover essential for Husserl – it is appropriate in this sense to interpret the judgment of Trân Duc Thao, who praised phenomenology for "the destruction of formalism within the very horizon of idealism" (Trân Duc Thao 1983). "In the horizon of idealism" certainly constitutes a restriction which we would share with Thao if we were to assess phenomenology; but we wish to refer to it only to nourish our reflection on science, and to instruct us.

Now, it seems that the logical analysis of language is ordinarily pursued in a perspective so purely syntactical that it cannot but lead to an exacerbated formalism, and consequently it fails to respond to the second part of our problem, that of the articulation of *logos* and of the concrete world. All logical research on the nature of forms in science – that is, all consistent epistemology – therefore should begin by re-examining received ideas respecting language and that is what I propose to do in the initial chapters of this book.

COORDINATION AND *SUBORDINATION* OF FORMS

1.11. If linguistic analysis appears to be quite indispensable for an interpretation of scientific forms, an essential precaution must govern its usage. The concern with systematization and unification of the instruments of objective thought naturally predisposes us to reject all pluralism of formal structures. One would then gladly expect to see scientific formalism deployed on a single level whose *coordinated* structures would be described by a sort of general grammar. Indeed, here, independent of any linguistic consideration, is the Kantian idea of a transcendental formalism. In the *Dissertation* of 1770 Kant had already affirmed that "the form of the world consists in the coordination and not the subordination of substances." The Leibnizian language here can be ignored, for the same wish for coordination continues after the critical revolution, when it is no longer a question of substances but of phenomena. Yet, it may be noted that it is perhaps just this Leibnizian inspiration, bound to a linguistic conception of rational thought that paradoxically governs the movement of transcendental analysis from which language is excluded: the order of the universe is coordination, just as, at first sight, is the order of words.

1.12. But all modern science belies this thesis, and besides, a better analysis of syntactical mechanisms leads to a view of language not only as a coordination, but also and essentially as a subordination of forms.

It could thus be said that in order to understand science, epistemology must make, rather curiously, a sort of return to Aristotelianism. By this I mean that one pays one's respects to the themes of reflection proposed by that philosopher of the hierarchization of structures, without it being in any way necessary to posit these structures as real parts of the world: It suffices that they appear as characterizing the moments of our laborious activity with respect to things.

1.13. Not only is scientific knowledge *discourse* about objects, it is also the elaboration of that discourse; and its articulation with perception requires that discourse about that very discourse be possible and that successive degrees of language become apparent. This succession need not be interpreted as an infinite regress, since science exists in fact, but only as a possibility, indeed an ever present necessity of taking the very instrument of scientific knowledge as an object of fruitful investigations, of moving from schemes to themes, and back again. If this is so, then it is the hierarchization, the subordination and the mobility of levels of construction that characterize scientific thought. The dream of a purely coordinative systematization of forms could only belong to a now bygone state of science.

1.14. Plainly, our assertion must be justified by the facts. The material that follows will furnish considerable justification borrowed from the data of the sciences of man, and these could also be found in an earlier work devoted to the methods of economics (*Méthodologie économique* (1955), for example, part I, chapter 1, and part II, chapters 2 and 3). But it would be sufficient to refer to the bulk of recent work in epistemology, generally devoted to the natural sciences, in order to find grounds for this claim. Whatever may otherwise be their philosophical orientation, epistemologists have, since the turn of the century, been led to emphasize the plurality of formal levels of scientific thought; in this way the project of a transcendental analysis has been profoundly modified. The description of the formalism of knowledge is now that of an *open* formalism, in which *error* has its place. The gropings of science are no longer simply accidents and episodes; they form an integral part of knowledge, which in no way consists in the progressive disclosure of a truth, all of a piece, but rather in self-instruction, that is, in the construction of explanatory fragments, in their demolition, in the increasingly improved awareness of their reciprocal meanings and in understanding their plurality.

A 'PTOLEMAIC' REVOLUTION

1.15. Under these circumstances a philosophy of the sciences could not rest exclusively on the exploration of transcendental subjectivity, as Kant or Husserl understood it. Certainly both contributed in an essential way to stating the problem of science, by defining the transcendental subject as the subject insofar as it focuses on an objective world from which it is inseparable. J. Vuillemin, commenting on the 'displacement' that intervenes in the critical philosophy when faith is substituted for knowledge, says that philosophy undoubtedly requires not a Copernican revolution, but a 'Ptolemaic' one: "Perhaps then these displacements would cease, and the philosopher might no longer need to replace knowledge with faith, for he will have begun by replacing the human *cogito* in a universe of gods with human labor in the world of men" (Vuillemin 1954, p. 306).

It is in fact a Ptolemaic revolution which the philosopher of science must carry out in passing from a doctrine of the *cogito* to a doctrine of the concept ... It must be recognized that pragmatism in its diverse forms has contributed in a certain way to this transformation; but it has done so only at the cost of an intolerable capitulation: the value of knowledge is dispersed by the randomness of the diverse and contradictory success that the techniques encounter, and in the end it is by means of a return to the most scandalous subjectivism that the books are balanced. But we must admit that this is too high a price to pay for an attempt to integrate action in knowledge.

Et propter vitam vivendi perdere causas.[1]

Nevertheless, it must be possible to maintain the objective value of a science even while accounting for both its history and its formal vocation. But I believe that to do so, it would be necessary to replace the transcendental analysis of the conditions of perception (that Kant extended into a meditation on rational mechanics), by an analysis of the conditions of *praxis*. The legacy of a transcendental philosophy stands out here in that an analysis of *conditions* rather than *facts* is required, but the requirements that we have encountered above could only be satisfied by the substitution of praxis for perception. Clearly, we need to understand precisely what is meant by *praxis*.

1.16. In the Marxist vocabulary the word is introduced in opposition to speculation, in order to designate those human activities which converge in the development of concrete social life. From this perspective, the perception of a subject that is radically detached from the determinations of its mode of life in terms of the natural and social environment, appears to be a dangerous abstraction. Perhaps one can admit this more easily if one adds – what present-day Marxists too often refuse to see – that this process of abstraction is itself an aspect of *praxis*, and plays an essential role within it. It would surely be very absurd to accuse the most abstract formalizing activity, such as that of contemporary mathematics or certain branches of physics, of degeneration, under the pretext that it systematically betrays the concrete vocation of man in his struggle with his environment. It would be no less unreasonable to proclaim that this activity alone, which brings us nearer to the speculative essence of our being, reveals to us the fundamentally subjective nature of knowledge. On the contrary I believe that it is necessary to attempt to understand that this formal speculation is really a moment of *praxis*, and to discover how it is articulated with those other approaches that constitute our activity. This should be the real task for a mature philosophy of the sciences, one that is concerned only with understanding, not with laying down the law or with bearing witness. As an enterprise outside of all dogmatic positions, this task is not without numerous difficulties. For the philosopher's first step must consist in an attempt at a comprehensive and yet sufficiently precise definition of this *praxis*, which envelopes but goes beyond scientific activity in every way. Thus, the risk of confusing a philosophical interpretation with strictly empirical research in history and sociology is considerable. Nevertheless, if sociology and history of culture must in fact provide some of the material for the philosophy of science, they cannot be substituted for it. While they permit a definition of the content of *praxis* at a given moment, philosophy, in taking the practice of science as its own object, attempts to analyze its structures and conditions, and not to offer a causal explanation of its history.

1.17. The epistemology of a scientific discipline thus presupposes a cognizance [*prise de conscience*] of the concrete nature of the scientific practice that corresponds to it, and not simply a speculation, however ingenious or penetrating, about theories. In the end, the fundamental question remains that of the *forms* of scientific thought. It is not a matter of describing an ideal formal universe, constitutive of all science, such as the system of Euclidean geometry or Newtonian mechanics could have appeared in Kant's reflections. In any case, if an epistemology of the human sciences troubles us more in terms of the as yet indeterminate character of their practices and the uncertainty of their results, it is by the same token far less likely to let us tumble into the illusion of a transcendental dogmatism. But nonetheless there remains the difficulty of attaining clarity in the confusion of often incoherent scientific practice, almost always carried to the excesses either of formalism or of empiricism. So, in my choice of examples from the diverse disciplines whose goal is objective knowledge of man, I shall try to make explicit the meaning and the role of formal thought in scientific practice.

1.18. In this regard, it makes sense to dissipate a prejudice that often clouds the appreciation of the value of formalism in the sciences. There is the usual confusion between formal thought and the work of mathematicians. If it is quite true in a sense that all effective scientific formalism tends towards a mathematical law, this still does not mean that the formalism is unavoidably reduced to the *usual* and *current* instruments of the geometricians. Yet is it not just this supposed reduction that the opponents of all formalization in the sciences of man are grappling with? It will be one of the aspects of our task to show formal thought at work in the human sciences, not only as a reduction of phenomena to calculi, but also as the invention of new structures, even, indeed, of an original mathematics. In any case, this formal creation appears to us not as disinterested play, but as a paradoxical and fruitful phase of *praxis*.

1.19. In the light of the preceding remarks (1.10), the plan that we propose to follow begins with an essay on language as the necessary mediation of scientific thought. The introduction of this theme constituted one of the most decisive steps in the progress of a philosophy of the sciences, initially formulated in the critical perspective. But this theme is far from being perfectly clear; the most recent ideas of linguists and logicians have been developing above all on quite distinct levels and it seems to me important that they be brought together. I shall try to draw out of this a conception of language that may clarify the chief problem of the relations between form and content in scientific knowledge.

Another chapter will then be devoted to the '*découpage*' of phenomena in the science of man. It is here that the first intervention of formal thought is noted. We shall examine two types of *découpage*, which in fact are interconnected and coordinated: one is, strictly speaking, 'formalist', the other 'operational'. Thus we shall see how the classical problem of 'definition' is transformed when it is envisaged as deriving from a *praxis*.

In the course of this segregation of concepts, the psychologist and the sociologist necessarily face the dilemma of the qualitative and the quantitative. We shall have to examine this dilemma, and this will bring us to the very heart of the debate between the exponents and the opponents of formalism.

By exposing and if possible dissipating certain prejudices, the discussion of the preceding theme will lead to the notion of axiomatization in the human sciences. Axiomatization is the subject of a chapter that will probably be found to be one of the most difficult, and most significant in this work. But the notion of axiomatic system, even interpreted as we shall attempt to do, cannot furnish the human sciences with the unique schema for their construction. In an opposition, which I hope to show to be complementary, there appears the problem of an understanding of the individual. It is to this problem that I have devoted the last chapter, with the thought that the most original aspect of a science of man will become particularly manifest in its future elaboration.

CHAPTER II

LANGUAGE AS A VEHICLE OF INFORMATION

῞Οτι, ὦ Σώκρατες, τῶν μὲν ἄλλων τεχνῶν περὶ χειρουργίας τε καὶ τοιαύτας πράξεις ὡς ἔπος εἰπεῖν πᾶσά ἐστιν ἡ ἐπιστήμη, τῆς δὲ ῥητορικῆς οὐδέν ἐστιν τοιοῦτον χειρούργημα, ἀλλὰ πᾶσα ἡ πρᾶξις καὶ ἡ κύρωσις διὰ λόγων ἐστίν (*Gorgias*, 450b).

[Because, Socrates, in the other arts all the knowledge, so to speak, is concerned with manual labor and activities of that kind; but in rhetoric there is no such manual operation. All its activity and effectiveness is through words (Plato 1955, p. 7).]

RHETORIC AND CONTENTS

2.1. Science is a discourse; he who passes over this condition in silence runs a great risk of losing his way completely. In fact if this aspect of science is ignored, nothing is left but a bundle of techniques, or more precisely, some series of badly connected gestures, effective perhaps, but static, proposing nothing to the mind but an exact and servile imitation, bearing in themselves no force of expansion and progress. Such is in fact the character of common knowledge, although it generally makes use of ordinary language to transmit itself, but without really exploiting it, and it is in this that common sense is to be distinguished from scientific knowledge. Common knowledge uses, so to speak, common language as a neutral vehicle; for scientific knowledge language is not only a vehicle between different minds, but also a mediator between one mind and its objects. Now, with such insistence on this linguistic aspect, one might be afraid of ending up straight away with a rhetorical conception of science. My intention, however, is to combat just this view. The rhetorical use of language is radically distinguished from its scientific usage in that the former is confined to a purely verbal universe. Thus, Plato puts into Gorgias' mouth the insolent and superb response that we have chosen for our epigraph: "... in the other arts, all the knowledge, so to speak, is concerned with manual labor and activities of that kind; but in rhetoric there is no such manual operation. All its activity and effectiveness is through words." The word 'rhetoric' thus rightly designates every self-sufficient utilization of language; the word 'science', on the contrary, contains a conjunction, in fact, one shrouded in mystery, of 'manual acts' and discourses.

2.2. It is interesting, parenthetically, to note the strange attitude of the sophist, revealing, in short, this divorce between word and act. Plato relates in the *Hippias Minor* that the most brilliant among the sophists took great pride in making his clothing, his ring, his shoes, by hand, when the art which he professed was precisely that of words, not of performance . . . It seems that the modern scientific spirit is characterized by the progressive discovery of a technique of language that permits one not only to transmit, but also to improve and modify effectively diverse material techniques. The importance of ancient mathematics as a prototype for science is explained at least partially because it constituted the first example of the formal expression of objective material manipulations. Geometry and arithmetic permit us to speak effectively of a class of fundamental operations carried out on things, operations which the adult man can truly learn to effect with ease only because he can describe them.

2.3. The sophistic orientation is utterly different. Naturally, it is not a question of discussing here a philosophy whose interest and merits I would be the last to belittle; I introduce its spectre only to illustrate better the relation between scientific understanding and language. Now it is beyond doubt that science begins and loses itself by means of language if verbal expression is closed on itself, and if science, which consists primarily of *speaking about* something, is absorbed in a rhetoric which ends up speaking of everything and nothing. In its rhetorical usage language is originally an instrument of action tried out on someone else, then the technique to which this usage gives rise generates an esthetic satisfaction, of such a sort that it becomes its own goal. If science is a discourse, it is obviously not so in the rhetorical sense that language occurs in it. In science words play a role of mediation, not only in the banal sense of a mediation between subjects – a mediation which the art of persuasion would turn into a manipulation of the other – but also in the sense of mediation between the subject and the world of objects that perception gives us and that scientific language helps to render manageable. These two functions of language in science have been most often confused, and it is to dissipate this confusion that I shall first apply myself.

2.4. Failure to interpret this double function carefully leads in effect either to seeing in science only a *persuasive discourse*, or to discovering in it only the *purely grammatical* construction of a pseudo-objective universe. This is the case with nominalist philosophies. Condillac's assertion that science is only a well-made language is a singularly ambiguous thesis. It signifies, first of all, that scientific language proceeds from a correct analysis of the data of experience, and that by means of its precise designations, it constitutes an adequate vehicle; but it also signifies that scientific language is governed by a syntax both rich and clear, which gives it a power suitable for coherent construction. Thus one easily slides into a grammatical conception of science, according to which the object ends up as nothing but the product of a syntactic activity whose fruitfulness surprises us. At the extreme limit of this development, philosophy is quite naturally confused with the role of policing scientific language, and with determining its rules of construction.

2.5. The evolution of neopositivism is in this respect quite instructive. Beginning with a 'logical empiricism' that postulates the position of 'protocol statements', which adequately designate primitive, atomic experiences, it was confronted by difficulties which led to the accentuation of its nominalist and idealist aspect, in spite of its strongly affirmed concern to reject all 'metaphysical' theses (Garaudy 1956). The data of atomic truths in perception appeared in fact to be very arbitrary; hence, Carnap and Neurath adopted the principle of 'tolerance', which frankly admitted the arbitrariness of initial hypotheses in all scientific language. The analysis of science was thus reduced to an analysis of syntax. But this radical position was untenable; Carnap was soon led to reintroduce the problem of the relation between language and experience: in the course of this later semantic phase he examined, in several penetrating works, not the effective articulation of the expression and contents, but the general conditions, of the designation of objects by symbols. But here language remains closed in on itself. One can agree with the neopositivists that science is a well-made language, but on the condition that it includes a concrete and not only a formal semantic activity. Yet at no time did the exponents of this philosophy ever really pose the semantic problem. For the thesis of protocol statements is nothing but a bar [to further analysis of this problem]: scientific data would be reduced to bundles of perception, and there is no longer an authentic *object* other than the very forms of language. By contrast, in my view, it is in disclosing and analyzing a specific linguistic activity at the very level of data from which science begins that we will easily be able to exploit the very real gains of neopositivist theory. The constitutive linguistic activity of science begins in focusing on *objects*, and not merely in the designation of sensations.

EPISTEMOLOGY, GENETIC PSYCHOLOGY AND AXIOMATIZATION

2.6. The philosophy of the sciences should then include a semantics, but not one taken in the exclusively formal sense intended by Carnap: an effective activity of *découpage* and designation of objects *a fortiori* merits the title of semantics. Its study could have profited greatly from the works of genetic psychology, such as those of Piaget. It is precisely because epistemology is in no way a psychology (which takes as its object the concrete individual being) that it needs to ask psychologists what is individual and actual knowledge. Piaget's research on the formation of logical thought in fact provides us with some very remarkable information about the characteristics of the *phenomenon* of scientific thought. The idea of 'forms of equilibria' of thought to designate the structures that the logician studies, as opposed to the process that the psychologist describes, seems to me to capture perfectly the relation between the logical and the psychological. It is a question neither of results somehow contingent on a process of empirical evolution, nor of some *a priori* ideal that causally orients psychic becoming. Piaget has shown us the imperfect systems which precede the system of logical manipulations. And what the psychologist must try to present as the moments that date from an origin and link to the ensemble of an activity through which the self is formed, remains for the epistemologist to examine as the structure taken in itself, or more precisely, considered as a mode of grasping a real object. But when Piaget employs the word 'axiomatization' to designate that situation of reciprocity that I am trying to describe, he perhaps needlessly hardens and deforms the position of the logician. It is quite true nevertheless that the task of epistemology, and of logic in general, is to uncover the structural organization of science. But when axiomatization – which is a part of science itself rather than of the philosophy of science – aims at determining coherent and finished structures, this normative character disappears in the epistemological enterprise. By the same token mathematical logic, which involves the search for the coherence mentioned above, detaches itself to that extent from epistemology in order to range itself among the sciences, on a par with mathematics, of which it is a specific branch.

2.7. The objective study of the forms of scientific thought is not, strictly speaking, an axiomatic enterprise, for one cannot look for coherence where it has never been. This search for coherence at all costs appears, on the contrary, to be within the scientific work itself, and the logician is not one who axiomatizes science, but one who examines how science tends to axiomatize itself. All the more as it is not without serious drawbacks that one can speak of axiomatization with respect to pre- or proto-scientific figures of thought, as Piaget does in his works on intelligence. The insurmountable difficulties which his notion of 'grouping' encounters are the consequence of this *μετάβασις έις ἄλλο γένος* [transfer to another genus]. In order to give an account of one stage of child intelligence, Piaget wanted to bring to light a specific figure of equilibrium or quasi-equilibrium, which he wished to treat axiomatically, as a kind of weak mathematical structure, a prototype of the 'group', a completed mathematical form. But to the degree that he axiomatizes, he succeeds only in characterizing in ambiguous terms structures more complex and less general than the group, *which are already mathematical structures*. And to the degree that he accurately describes proto-logical structures, his axiomatization fails.

CRITIQUE OF THE NOTION OF 'GROUPING'[1] AS A FORM OF LOGICAL THOUGHT

2.8. We know that the structure of 'grouping' introduced by Piaget is characterized by the *intensive* organization of classes of objects of any sort. By 'intensive' we are to understand "quantitative relations, which include exclusively the inequality of the part (class A) and the whole (class B), either $A < B$, or the identity ($A = A$ and $B = B$) without consideration of the quantitative relation between one part (class A) and the other parts disjoined from A (classes A'), belonging to the same whole (class B), or between these parts (classes A or A') and other parts (class B', etc.) belonging to other totalities" (Piaget 1949, p. 72).

On the other hand, extensive relations, specifically mathematical ones, obtain between all disjoint classes of one and the same whole or of different totalities.

Grouping is then defined as a type of structure governing the relations of classes of a system, whose prototype is apparently the tree-like schema of class inclusions

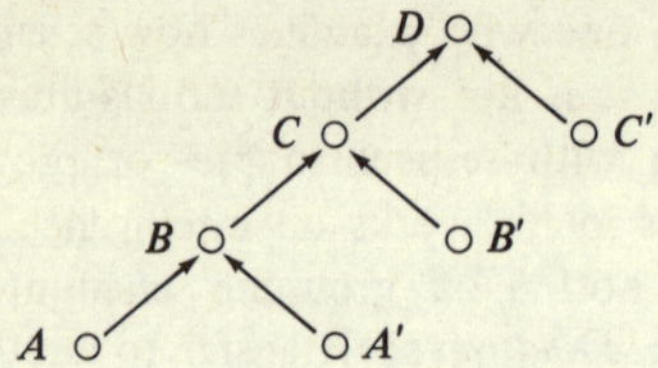

$$A \subset B \subset C \subset D$$
$$A' \subset B$$
$$B' \subset C$$
$$C' \subset D$$

Piaget defines this scheme in terms of the properties of '*operations*' (of union of classes) which it permits:

(1) A 'direct operation' on two complementary classes, such as C and C', B and B', resulting in an including class, D or C.

(2) An 'inverse operation' of exclusion on two included classes such as D and C which results in the complementary class C'.

(3) An 'identity operation', the union of the *empty* class with any other class, which is thereby unaltered.

(4) Certain 'special identity operations' which consist in the union of an including class such as D and one of its sub-classes such as C, or even the class D itself:

$$D + C = D \qquad D + D = D.$$ [2]

(5) The above 'operations' possess a property of limited associativity because of the necessity of operating on contiguous (i.e., included or complementary) classes (Piaget 1949, p. 98).

This type of structure is contrasted with a *Group*, characterized essentially by a greater operational freedom: its operations are carried out on any two classes of the set in question, and not only on two contiguous classes. Grouping is also differentiated from Group, in that it involves 'special identity operations'.

While willingly acknowledging the insufficiently general and 'inelegant' character of the notion of Grouping, Piaget insists precisely on the fact that it would be suited to systems of operations of the logic of parts and wholes in its premathematical form (Piaget 1949, p. 99), and that it is consequently important to avoid reducing it to more satisfying but later mathematical structures.

2.9. Let us examine this notion more closely. In the first place it will be seen that the presence of 'special identity operations' is in fact incompatible with Group structure. It is trivially demonstrated that, in a group, there could be no other *idempotent* elements (such as $A + A = A$) besides the unit element.

The other distinctive trait, namely that the operations of Grouping are not in general defined between any classes whatsoever, is certainly much more significant. In the course of numerous experimental observations, Piaget has shown that the thought of the child proceeds first through elective and limited combinations of the elements it structures. It is in this sense that the objects of this thought can be called 'qualitative,' since they are differentiated and non-interchangeable.

2.10. Nevertheless, this opposition between Grouping and Group leaves us with a certain uneasiness which at no time have Piaget's analyses ever explained. Mathematical and pre-mathematical structures are defined by him in theory as organizations of *operations* on *elements*, which are classes. Now a serious ambiguity hovers over the very notion of structure as Piaget uses it. In his works on the psychology of intelligence he makes reference repeatedly to the concept of 'operation', and seems to define Grouping itself as a system of 'operations'; this idea, as he uses it, is confused. In fact, a structure like Group or Grouping can be considered from two points of view, which it is important for the logician to distinguish. In the first place one can consider a structure as a system of '*operations*', or more precisely of *transformations*', attention then being paid not to the entities to which the transformations are applied, but to their properties of mutual composition, that is, their interconnected applications. This is the case, for example, in geometry, with the Group of translations along an axis defined by their direction and their amplitude. The successive application of two translations yields as a result a translation whose amplitude is the algebraic sum of two others, and for every translation there is always an associated translation – its inverse – which, combined with the first, cancels it . . . . It seems that it is this sort of thing that Piaget has in mind when he speaks of 'operations'. We should therefore note the 'operations' of the Grouping of classes by the symbol designating the class to be joined, followed by the addition sign ($A+$), since the transformation which consists in the union of A with any class whatever is then essentially distinct from the transformation which consists in its union with a class B ($B+$), and the name of the class, A or B, loses its substantive sense through its incorporation into the sign of the operator. (Like the number which represents the amplitude of a translation.)

Note how inadequate this view of transformations is for Grouping, since a class generates an operator only if the operand is a contiguous class. In the final analysis, a symbolization of the type: A_+ to designate the transformation from the indeterminate class X to the class $A \cup X$, is impractical, since it makes sense only for X's that are contiguous to A. It would doubtless be better to retain the notation: $X + A$; but then one no longer knows whether one is speaking about the *transformation* as defined above – which Piaget calls 'operations' – or about the abstract *operation* symbolized by the +, considered independently of the elements X and A upon which it operates. In order to be applicable here, the notion of transformation should be definable without reference to its subject of application.

The notion of *operation* in the mathematician's [3] sense is based on another conception of structures called abstract. In this case one considers a set of *entities* provided with a *law of composition*; it is the properties of this law that must characterize the structure, Group or Grouping. From this perspective, the law of composition, i.e., the unique operation of Grouping, is the union (symbolized by '+') of two classes, and is defined only for two contiguous classes. Thus one can no longer speak of 'operations' in the plural, but only of entities, and there are no more identity, general or special operations, but a *neutral* entity or *zero* – the empty class, which, while composed with every other class, leaves it unaltered.

2.11. In the case of Grouping the difficulty derives from a vacillation between these two perspectives, which masks the impossibility of a coherent interpretation.

In fact, when Piaget defines the notion of Grouping, he speaks in the singular, of a '*direct operation*' (and its inverse) which is the union of one class with its complement under their including class. Since this then involves *one* operation and not a multiplicity of operations corresponding respectively to each class, Grouping is to be understood as a system of elements (classes), provided with a law of composition, which consists in joining two of the (contiguous) classes.

But the same no longer holds for the notion of 'identity operation' which imposes another perspective on us. A coherent conception would require that one speak, on the one hand, about a neutral entity, the empty class, which, united with *no matter what* class, leaves it unchanged, and on the other hand, about the *property of certain* particular *pairs*, whose combination leaves unchanged one of their elements. If one insists on speaking about 'special identity operations' one must adopt the language of transformations.

In the abstract language of the operation, the phenomenon of special identities corresponds to a property of the *elements* and not of the operation itself. Thus, it can be said that the operation of special identity breaks down, for if one is considering a Grouping of transformations, then the specification of the operative element could not enter into the definition of each of them, but only into that of the operator; there will be an operation A_+, an operation B_+, etc. ... and the characteristic "A_+ applied to A yields A" is reproduced for *all* the operations.

If, on the contrary, one envisages a Grouping of elements under a law of unique composition +, one could not speak of 'identity operations' either, but only of a specific property of the previously mentioned law, i.e., that which makes a result of the type A or D correspond to a pair of the type (A, A) or (D, C).

Thus, the word 'operation' in Piaget's definition is taken sometimes in the normal sense of the law of composition (Piaget 1949, properties 1 and 2, p. 97), sometimes in the sense of transformation (property 3), sometimes in a sense incompatible with the two preceding ones. The fact is that the notion of Grouping is in itself complex and ambiguous. It involves the idea of a system of operations which is *superimposed on an ordered structure* of the objects on which Grouping is defined (the class inclusion, or more generally a partially ordered structure in the mathematician's sense). Undoubtedly Piaget's idea was precisely to show the operative foundation of this ordered structure, and he believed it necessary to introduce to this end his notion of Grouping, with the express purpose of constituting a type of 'intensive' structure. In fact mathematicians have already axiomatized this operative foundation of class inclusion by constructing the notion of *Lattice*, and without doubt it would have been fruitful to start from this structure with two operations (corresponding here to union and intersection) in order to explore the first attempts at the classification of the thought of the child.

2.12. This long critical digression, which has permitted us incidentally to examine a weakness in one of the most original and vigorous epistemological constructions of our epoch, was aimed chiefly at showing that it is not the task of the logician to axiomatize the structures of scientific thought. The intuition of proto-logical structure developed by Piaget is rich in meaning but on the condition, I think, that we see here not axiomatizable systems, but the premises of a development of thought which leads precisely to axiomatized structures, real forms of equilibrium of rational thought, i.e., to mathematics understood in the broadest sense. The movement of axiomatization is given in science itself, and the 'qualitative,' 'intensive' thought of Grouping is thereby rendered scientific, that is, susceptible of axiomatization.

ORDINARY LANGUAGE AND FORMALIZED LANGUAGE

2.13. We must return to language. Is it not the case that this critique we have made of Piaget's genetic psychology of intelligence can be equally well directed against his total neglect of the linguistic element in the formation of scientific thought? Naturally, it is not for us to discuss this deficiency from a psychological point of view. Piaget thought that he would be able to justify it by claiming that the acquisition of language accompanies the formation of logical thought, but does not determine it. Even if we limit ourselves to the consideration of logical thought as a set of works and a methodical system of operations, we still cannot avoid stating how much the neglect of the linguistic aspect handicaps Piaget's epistemology. A comparative study of the systems of thought described by him as proto-logical (for example in *The Child's Conception of Number* (Piaget 1952)) leads quite naturally to bringing to light the characteristic role of linguistic elaboration in the transition from vague experience to science. The structures of the child's pre-scientific thought, as he describes them, are either in some sense prelinguistic structures, or, on the contrary, reveal a use of language which classes it as a thing. In *The Child's Conception of Number* Piaget shows children a certain quantity of colored liquid which he, standing in front of them, pours into one flask after another, each of a different shape. When questioned, the children judge the quantity of liquid according to the shape of the receptacle and the level the liquid attained; the manipulation of the liquid before their eyes or by the children themselves was insufficient to give them the sense of the invariance of a quantity. When children move beyond this stage, it is no longer experience alone which convinces them, but *spoken experience*. The psychologist must put before their eyes, so to speak, not the facts, but contradictory *expressions*, so that they arrive at an understanding which, once grasped, anticipates perception. Thus, it is to a process of linguistic construction of scientific thought that Piaget's fine and numerous experiments testify. And the reversibility of operations, which Piaget rightly designates as the fundamental character of logical structuration, is grasped right in spoken experience, rather than in perceived experience. In the end, the fundamental problem of scientific knowledge lies neither in the development of a language, nor in the variation of experiences, but contrary to Gorgias' rhetoric, in the collaboration of a linguistic expression and a manipulation.

We propose to examine this collaboration at certain points in the domain of the human sciences. But first we should make clear the very notion of linguistic formulation.

2.14. If we admit that all scientific behavior surrounds a fact of language, it is nevertheless certain that not all linguistic behavior is necessarily scientific. The structures of scientific thought depend on an aspect and a special usage of language. This postulate has been implicitly adopted by all the recent epistemologies which have introduced linguistic considerations as essential. But the consequence that they draw from this postulate is that only the formal, syntactical aspect of language operates in this context. The consequence of this was the flourishing of logical syntaxes. We noted above (2.5) that the very evolution of neopositivism revealed a disquiet in the face of this grammatical reduction of the structures of thought.

It seems therefore necessary, in my view, to return to an examination of common language, as it is actually employed outside of scientific formulations, if one wants to really recognize all the functions which have been masked by the very real predominance of the syntactic element. Logicians in effect describe scientific language as an empty structure, as if the *raison d'être* of a language were simply a certain grammatical perfection, and not communication among men. However, as Wittgenstein so forcefully noted, in ordinary language "we are not *striving after* an ideal, as if our ordinary vague sentences had not yet got a quite unexceptional sense, and a perfect language awaited construction by us" (Wittgenstein 1968, § 98, p. 45^e). Even leaving aside the emotional and mimetic aspect of language, it is necessary to recognize in effect that language is ordered essentially for communicating information, and that the very notion of syntax, which assumes such importance in the philosophies of scientific expression, is somehow secondary. Now, whatever the syntactic hypertrophy of 'formalized' languages constructed for the use of mathematicians and physicists, it is nevertheless true that they retain their original function as *vehicle*. The notion of formal thought could only be altered by the extraordinary neglect of this truism.

PURE INFORMATIONAL LANGUAGE

2.15. It is remarkable that it should be technical, and even strictly pragmatic considerations which today suggest this revision of the philosophy of forms.

In fact it is the research of telecommunications engineers which contributed to giving a scientific status to the vague notion of information, and at the same time brought to light this fundamental aspect of language. What in fact is a linguistic expression, such as, 'the kitten is dead'? It is a *discrete linear sequence* of elements chosen from a *dictionary*, previously known to its users, whose choices are *limited by syntactical rules*. Since we are concerned with ordinary languages only in order to understand better the mechanism of scientific languages, we may legitimately leave out of account the emotive resonances, and the supplementary information which the diction – or the typography, the layout – superimposes on the fundamental information. These are all things which a integral philosophy of language cannot neglect, but which scientific thought carefully avoids. Besides, it is just these *appoggiaturae* of expression which make people often pretend to consider the fine arts and mimetic sign languages as true languages. But the means of the musician, the painter or the mime, even when expressed with a supreme richness, must nevertheless be considered as infralinguistic unless we extend the notion of language to every sort of expression. Common sense obscurely recognizes this specificity of language in its refusal to accord speech to higher animals even though they do express themselves.

Linguistic activity, thus described, consists in a series of choices among given vocables. It is important to understand that, strictly speaking, a syntax is not necessarily required for the existence of a language. It is perfectly possible to conceive a linguistic activity as nothing but the unconstrained juxtaposition of the vocables drawn from a dictionary, which thus compose a message. Doubtless, this is the earliest mode in which language appears in the child, and towards which language tends when it degenerates. A deficient language, certainly – and we shall soon see what syntax offers it. But then, it is the emotive and mimicking *appoggiature* that tend to make up for such a language's deficiencies. In it, each element naturally finds itself autonomous and isolated. The structural interconnection is furnished from without by intonation, or its graphic substitutes.

2.16. Under these conditions, a pure linguistic sequence is a vehicle of information simply by successive decisions that the sequence effects by producing one 'word' from the dictionary, rather than another. It is thus possible, in this rudimentary but essential schema, to define and even quantify the information, if it is admitted, as is reasonable, that the number of entries in the dictionary being used is finite and determined. This finite character of the dictionary is, in fact, fundamental. Indeed, common languages require a very impoverished dictionary. At the phonemic level, spoken languages hardly use more than 15 to 75 distinct units. Common written languages, considered as autonomous languages, have a still smaller number of signs. But the richness of a language stems from the combinatorial possibilities of its signs: given 15 phonemes one can in principle construct 15^n 'words' each containing n phonemes. Syntactic rules (we are dealing, of course, with phonetic rules) obviously restrict this potential richness, but they accordingly diminish the chances of distortion in messages (cf. 2.18).

For greater convenience, let us assume a very impoverished dictionary containing only two signs. A message of 4 signs, for example, will express a series of 4 independent choices between the two terms of the dictionary. If nothing *a priori* influences or limits the choice of the sender, $2^4 = 16$ distinct sequences of 4 signs are equally possible. The statement of a determinate message thus dissipates the uncertainty of our expectation in the presence of 16 virtually expected statements. If the base dictionary had included three signs instead of two, the number of possible messages would have been $3^4 = 81$, and our previous uncertainty would have been increased. The information carried by the message in the universe of 3 signs can thus legitimately be considered to be greater than that transmitted in the universe of two signs; one can even agree that the number of messages, which are *a priori* possible, *measures*[5] the volume of our uncertainty, and that every arbitrarily increasing function of this number will conveniently measure the information carried by a determinate message that dissipates this uncertainty. The binary logarithm of this number[6] has been chosen in order to reduce all informational language to the simplest language, involving only yes and no.

2.17. Thus we see how a strictly semantic structure is introduced into language considered as a vehicle of information. Some remarkable analogies, which we need not enter into here, have led to borrowing the physicist's term *entropy* to designate this amount of information, or if one prefers, to measure the certainty of expectation dissipated by a message of a given length. So, the linguistic expression appears as an instrument of communication, and from this point of view, which could be regarded as strictly material, formal properties, very different from syntactic properties, emerge. Thus, in particular, instead of admitting *a priori* the equiprobability of the occurrence of all the elements of the dictionary, one will be able, in accordance with what is the rule in ordinary languages, to consider that each sign has its own probability of appearance, and that the entropy of a message must be calculated by taking into account the fact that a message composed of 'unusual' signs carries more information. The rather technical character of a mathematical theory of information must not cause us to lose sight of the fact that here lies a hitherto totally neglected conception of language, and that the formal element does not make its first appearance at the level of syntax, but is already present at the level of the dictionary. The linguistic construction of science demands examination beyond the structuring that this syntax imposes in the very designation and choice of the signifying elements of discourse.

2.18. Moreover, communication theory permits us to renew in one sense the classical conception of syntax. In the message cited at the outset: 'the kitten is dead', the syntactic rules require that the first word – an 'article' – can only be followed by a word of a certain category, i.e., 'adjective' or 'substantive'. Generally speaking, by reducing the freedom of choice of successive vocables, the grammatical apparatus restricts the indetermination of our expectations. From the point of view of information the result of this is that each word of a message carries less information than it would in a language without grammatical structure. This all happens, somehow, as if one used more words for the same quality of information than would be strictly necessary, or although the information of each word overlaps that of the others. In a broad sense the function of syntactic rules is, from this point of view, comparable to the repetition of certain signs in a telephone message: "Arsenic: spelled *A*natole, *R*aoul, *S*uzanne ... " We say that a syntactically structured language thereby possesses a *redundancy*, a concept which communication theory enables us naturally to quantify on the basis of the notion of information, and which characterizes, not an individual message, but a language in general, since the incidence of syntactical rules can obviously be considered only as an average and in a random fashion. Semantics and syntax are thus to be interpreted as distinct formal contributions inherent in all linguistic expression. Their superposition and their complementarity will be better understood if one imagines two extreme and fictitious types of language. One, totally asyntactic, in which every vocable is autonomous; any juxtaposition whatsoever of vocables is possible, redundancy is nil, and the information carried by a message is a simply additive function of its length; the other, completely interconnected, of such a sort that the appearance of an initial vocable syntactically determines the following vocable, and all information is, in the final analysis, entirely syntactical.

It seems to us, then, that it is characteristic of the informational point of view to clarify the roles of semantic and syntactic forms introduced by language into scientific thought, by ruling out any temptation of a metaphysical interpretation of structures. But this opposition of semantics and syntax should be rendered more explicit in order to enable us to better understand the formalist aspect of knowledge and to dissipate misunderstandings of it.

SEMANTICS AND SYNTAX

2.19. The linguistic sign, in every ordinary language, is a bearer of complex senses. In general it is capable of carrying out three sorts of roles. One is to render psychological attitudes, emotions or actions, which it tends to induce in others or from which it is supposed to emanate in the speaking subject. The word – or the sentence – remains in this sense at an ordinary infralinguistic level; it is certainly already intentional, but it remains bound to the living experience [*vécu*] of the subject which it expresses.

Secondly, the word refers to the 'objects' which it designates, and from which it is then detached. The strict notion of semantics in principle refers to this function of the word as name, and the modern theory of information takes this as its point of departure.

Finally, the word is a possible bearer of a syntactic sense, that is, it may refer to structural rules that concern it insofar as it is a sign. It is important to note that the redundancy thus introduced into a language is, as such, a property not of the objects designated but of the language itself. The scheme of the Aristotelian apophantic '*S* is *P*', to which each word refers insofar as it is either a subject or a predicate, is quite obviously a property of a certain language. We know that the voluntary or spontaneous confusion between objective structure and syntactic structure has often been condemned and this denunciation has been a commonplace of neopositivism. Nevertheless, we do not dwell enough on the essential positive fact, that all fruitful scientific thought is precisely an effort to construct a language whose syntax authentically contains the power to inform us of the objective relations of phenomena. But, far from appearing as a purified and perfect scheme of a world of images, the linguistic universe of science is the product and the instrument of an activity carried out on the perceived world. To say that it imitates better and better the structure of things is superfluous, *for without it there is, strictly speaking, no structure*. The idea of articulated structure is originally linguistic, which is not to be understood in the nominalist sense; the world is still objective structure plus language.

2.20. We must make this statement clearer by showing that the distinction between a syntactic function and a semantic function, between the reference to 'objects' and the reference to linguistic structures, is essentially relative and changeable. If this is the case, one can dismiss, back to back, a nominalist philosophy of knowledge, which makes the syntactic aspect an absolute, – and a crudely realist philosophy, which requires that every element of language always be *the name of something*. Since our principal object here is not a logical study of language, we can limit ourselves to just an examination of some of the linguistic levels in use, in order to show the interplay of these two functions (cf. Granger 1957).

First of all, consider *writing*. Taken in itself, and not as the transcription of a spoken language, writing is indeed a language corresponding to the definition given above (2.15). It involves graphic signs, whose usage is governed by the rules of its own syntax, which limit our freedom in the composition of a sequence. It is in fact difficult to make a clear-cut distinction between what belongs to this 'orthography' *stricto sensu*, and what is only the graphic reflection of the rules of the language for which the writing serves as a transcription. This is because writing is in reality thought of as a code and not as a language. Nevertheless, there are certain rules of structure which properly belong to it; for example, it is not permissible to write several semi-colons in a series, or again, capital letters in the middle of a series of lower-case letters . . . To this rudimentary syntax a richer semantics is naturally joined. Graphic signs designate sounds. The rules of designation of sounds are obviously not independent of the spoken language which writing transcribes, as we see in the example of our own writing system which transcribes a large number of diverse languages; nevertheless it is not at all absurd to think that writing, as an autonomous system, is not arranged for the transcription of another language, but directly for communication, as, in a sense, is the case with musical notation. Perhaps the most approximate realization of this graphic language would probably be offered by literary Chinese. Its mode of construction of ideographic and ideophonic symbols presents the rules of its own syntax which are not at all reflected in the spoken language. More analytic and richer than sound, the ideogram is said to permit abridgments of expression, rapprochements and symmetries which are valid only in the graphic space. The written language of the Chinese thus appears, in many respects, as a mode of expression that is very independent of oral language (cf. Margouliès 1943).

Let us now examine spoken language. If we want to try to consider spoken language in itself, we see that the signs out of which it is composed are sounds, or more exactly, *phonemes*: that is, classes or relations of sounds capable of being physically differentiated without ceasing to have the same significative value. (Just like graphic signs drawn by different hands.) The phonemes of a language are not freely associated, and the syntax that governs them must not be confused at all with the grammar. This syntax is constituted rather by the set of laws of the phonetic system, by virtue of which, for example, this group of consonants is inadmissible, or that vowel calls forth in its neighborhood a vowel of related tonality. Thus it appears that the transcendent objects designated by the signs of writing themselves remain signs, the set of which is provided with a structure; the *content* of the first level becomes the support of the *form* of the second; the semantic element becomes the syntactic element when the transition from writing to the spoken language occurs, a transmutation which is very exactly expressed by the transition from sound to phoneme. In my view, this is the general principle of the hierarchization of languages, which makes clearly apparent the relativity of the semantic and syntactic points of view in linguistic activity. All the more is this the case in the rigorous scientific execution of this activity.

2.21. Mathematics demonstrates this principle in its extreme form, but all science takes a hand in it. In fact, mathematics even appears to reduce itself to a pure language, because the syntactic element devours the semantic: mathematical signs no longer refer to objects transcending language, but to the laws of their own structure. The signification of so simple a symbol

	Syntactical forms	Semantic content
Writing	Graphic rules	Sounds
Oral language	Phonetic rules	Words
Language	Grammar	Ideas
Formalized language	Logic	Relations

as a number is by no means reduced to the designation of a determined object, of a collection, taken in itself. It involves a method of comparison of collections with each other, as well as the fundamental schemata of recurrence which through their usage define the series of numerical symbols.

From this to the idealist interpretation of mathematical entities is but one step, which I believe we must avoid taking. For this syntactical reduction, so radical in mathematics, always starts from a pre-symbolic datum which includes at the same time phenomena perceived as things and our own acts, their limits and their power. Starting from effective manipulations, language multiplies our possibilities of action through the intermediary of an imaginary world. Formal thought appears to me to consist essentially, from this point onwards, in the construction of a more and more precise syntax based on a more primitive semantics, which corresponds to names of things as yet poorly defined. It is from these syntactic systems thematized as quasi-objects that the process then starts off again. Thus, language, at the heart of objectivity, permits the differentiation of distinct levels and layers, that neither perception, nor a technique of immediate manipulation could do. How is this linguistic construction effected in the human sciences, to what extent is it legitimate, what can it offer for the effective understanding of phenomena? These are the questions we propose to examine. However, the informational conception of language that we have assumed leads us only to the threshold of formal thought. It still remains for us to bring to light certain comprehensive considerations about artificial symbolisms which necessarily extend ordinary language in the exercise of scientific thought, as we have just sketched it.

CHAPTER III

SCIENTIFIC LANGUAGES AND FORMALISMS

> " . . . the patriotic archbishop of Canterbury, found it advisable – " "Found *what*?" said the Duck. "Found *it*", the Mouse replied rather crossly: "of course you know what "it" means." "I know what "it" means well enough, when *I* find a thing", said the Duck: "it's generally a frog or a worm. The question is, what did the archbishop find?"
>
> *Alice's Adventures in Wonderland*
> (Carroll 1963, p. 26)

THE 'MIXED' LANGUAGE OF SCIENCE

3.1. If scientific language must not be detached in any way from the general conditions of the exercise of language without taking precautions, it nevertheless constitutes an original and highly differentiated aspect of this exercise. As I insisted above (2.13 and 2.14) it must be viewed as a vehicle of information; on the other hand we should now make clear its resources and special functions, in order to understand better the role and the nature of forms in scientific thought.

Logicians, who most often are interested in the language of mathematics – or even, more radically, in mathematics as language – have, by eliminating certain secondary qualities from scientific discourse, allowed its concrete aspect as instrument to escape notice. They pretend to see in it only a wholly constructed symbolism, whose artificial character is emphasized by those who imagine contrasting it to 'natural' languages. This is the thesis of the linguist Leonard Bloomfield, one of those who have most directly and most effectively considered the question. By language in the strict sense, he understands 'natural' *spoken* language; formal dialects, such as the systems of symbolic logic, although they possess a linguistic status, could not be independent languages (Bloomfield 1939, § 27). This is because in the final analysis they can never be anything for him but modes of writing, and the original linguistic activity is phonetic and not graphic. Undoubtedly, this conception of one linguist corresponds to a real subordination, both biological and social between the modalities of language. But it seems to me that the written character of scientific discourse is the object of a grave confusion in this regard. In fact the writing of ordinary languages is in most cases only a transcription, a secondary code, and if it has been possible for us to treat it repeatedly as an example of language, it is on the express – and artificial – condition of considering it in itself, independently of the spoken language to which it is naturally subordinated. But what was then only an artifice of analysis, aimed at a better understanding of the nature of linguistic activity through enabling us to avoid the prejudices involved in the too constant usage of speech, becomes the most adequate attitude toward scientific symbolisms. In scientific expression, writing is no longer a code: it is the very stuff of language.

3.2. We shall return shortly to the consequences of this *essentially* graphic character of the language of the sciences. But first I want to describe more explicitly this language as it presents itself, in order to discover the reasons for the misunderstandings that were just denounced. If we open a physics text, or one in mathematics or chemistry, we discover that the author makes use most often of a mixed language, in which he alternates sentences of the vernacular and formulae of a specific symbolism. Pragmatically, it is quite clear that ordinary language plays role as indispensable vehicle. By means of this language experiments are intuitively described, the rules for the employment of the symbolism are indicated, and in an even more precise way, the movements of a logical syntax are set forth, which permits the interconnection of segments of the formal language. Thus, it might appear at first glance that ordinary language, in essence oral, can, at a pinch, suffice for scientific discourse, the specific formalism of the physicist or chemist serving here only as a particularly concise and also precise abbreviation for vaguer and less compendious natural expressions. In fact every science that is only slightly advanced tends to a formal character, and the proportion of ordinary to formal language in each domain of science, indeed even in the works of each scientist, determines a style of scientific thought that is not without analogies in the styles of literary expression. This is not the place to dwell on the search, however captivating, for an esthetic, so understood, of scientific language. Nevertheless, there would be much to gain from such a study, in particular when applied to mathematical works. Let me limit myself to indicating, from this point of view, the 'pragmatic'[1] character, in Carnap's sense, of the diverse forms of equilibrium between vernacular language and formal language in scientific works.

But I believe it would be to mistake the true nature of this half-formal, half-common language if we see in it simply the reflection of a temperament or an epoch, and postulate, without further ado, its reducibility *de jure* to a pure formalism. The linguistic process of science seems to me essentially ambiguous: for if science is not at any moment of its history a completely formalized discourse, it is not to be confused with ordinary discourse either. Insofar as it is thought in action, it can only be represented as an *attempt* to formalize, *commented* on by the interpreter in a non-formal language. Total formalization never appears as anything more than at the horizon of scientific thought, and we can say that the collaboration of the two languages is a transcendental feature of science, that is, a feature dependent on the very conditions of the apprehension of an object.

THE FORMATION OF THE LANGUAGE OF CHEMISTRY

3.3. In order to understand the diverse aspects of this enterprise in a concrete fashion, let us briefly examine the history of the language of the chemists. Naturally, its origin can be only artificially detached from that of experimental manipulations; in any case I am not claiming to sketch a history of chemical language independently of a history of the whole of this science. The reader will restore the stages of experimental discovery which are continuously understood here. In limiting myself to isolating successive types of a linguistic activity, I in no way mean to present them as the well-springs of the progress of science. But since my purpose is only to elucidate a too often neglected aspect of scientific understanding, I will deliberately take this point of view which I know to be partial.

The first trace of a [chemical] notation is to be found in the alchemical manuscripts copied in Venice during the tenth or eleventh century, from which Marcelin Berthelot made a scholarly edition (Berthelot 1888). The signs used are at first essentially substantives, and language is reduced to its semantic function. These are the symbols for the seven metals, which are assimilated to the symbols of the seven planets: gold-sun (☉), silver-moon (☾), iron-Mars (♂), etc ... What the symbolism brings to the understanding is precisely that overdetermination of concepts that makes poetry possible. The magical and mystical intention is obvious, and the texts of the great Greek alchemists naturally confirm this interpretation. A formula such as the 'crayfish' famous among scholars is no doubt a kind of hieroglyphic memorandum rather than an objective proposition, or even a clearly defined recipe. 'O *νοήσαι μακάριος* [the blessed thinking] says Zosimos, at the end of one of his writings. Technical language, far from diverging here from the conditions of ordinary language, pushes its consequences to the extreme, and it mysteriously suggests a vision of the universe.

Nevertheless, among the alchemical signs, certain original traits are already apparent, as the rudiments of a scientific expression. Thus, on the one hand the signs are introduced for certain operations and instruments of the alchemists: One designates the crucible (≏), another the operation of pulverization (୭), for example.

In addition, a rudimentary syntax is sketched, with rules for the formation of compound signs. The symbol for gold leaf (°□) is composed out of the sign for gold while the symbol for silver leaf is based on that of silver (☾□).

But the role of such a language, reduced to an almost uniquely categorematic language, nevertheless remains secondary, a magical or charlatanesque ornament of ordinary language – otherwise quite emphatic – in which the obscure or uncertain messages of the alchemist are delivered.[2]

3.4. The appearance of a richer combination of signs and the progress which it produces in language mark the transition to a more elaborate experimental science, and the formation of positive theoretical hypotheses. One must await the age of Lavoisier to see linguistic development attain this stage, though still very timidly.

In a 'Mémoire sur la dissolution des métaux dans les acides' (Lavoisier 1782, p. 492), Lavoisier made explicit usage of a formulation which he himself was eager to provide as merely "a simple notation, the object of which is to ease the operations of the mind." However, it is here that the first attempts at a modern chemical language can be recognized.

Here are the essentials of his approach. First he chose symbols for water (▽), 'nitrous air' (△+), the 'oxygen principle' (⊕), and iron (♂: this is the old sign that the alchemists used.) Using these symbols he then wrote down the first state of the reaction that he was studying, oxydation of iron by diluted nitric acid:

(♂) + (▽) + (⊕ + △+).

The result of the reaction is then expressed as follows:[3]

(♂ + ⊕) + (▽) + (⊕ + △+).

But not content with this qualitative transcription, he added before each symbol a co-efficient of weight whose constancy the experiments had suggested.

(a♂ + α⊕) + (b▽) + (c⊕ − α⊕ + d🜂).

The symbolism here is arranged for a precise expression of the experimental conditions, and is governed by a fundamental rule of construction, to the effect that the symbols must remain the same in the two formulas, corresponding to the two states of the compounds present; to this rule is added a law of the conservation of the respective co-efficients of weight. Clearly, this is only a trial, and Lavoisier himself insisted on its fragmentary and provisional character by stressing with a truly brilliant perspicacity the ever present possibility of an experimental discovery which would require a splitting up of these symbols into simpler ones: "Doubtless, one day we will be able to decompose nitrous air, and perhaps the oxygen principle itself, and we shall be forced to replace them in these formulas with the expression of the principles which compose them . . ." (Lavoisier 1782, p. 524). Nevertheless, because of this very limitation, Lavoisier's formulae are the outline of a scientific language provided with precise syntactical rules and referring to well-determined manipulations. Fierz-David, who alludes to this 'Mémoire' (Fierz-David 1945) but does not cite the text mentioned above, is thus wrong in speaking of it as an "apothecary notation rather than a chemical one."

3.5. This concern with an *articulated* language, which we have just seen make a timid appearance, shows up in the same period in the systematic efforts of two French chemists, Adet and Hassenfratz, who explicitly introduced complex symbols for compound substances. The sign for hydrogen being ɔ and that of oxygen –, water was to be represented by the symbol ɔ̄ (cf. Kopp 1931, II, p. 424).

If one connects the idea of quantitative representation brought to light by Lavoisier to the idea of a combination of signs, one comes quite close to the modern language of chemistry such as Berzelius established at the beginning of the nineteenth century. The atomic hypothesis having already taken shape through the work of Avogadro and Dalton (who himself had invented a symbolism of the same type), notation could develop into a real language, with a syntax both simple and rich.

The Swedish chemist, making a clean break with the magic symbolism of the ancients, proposed to designate each element by the initial of its Latin name, and to make it represent one *atom* of the element, that is, a determinate mass characteristic of each of them. Compounds were to be represented by the combination of several simple symbols, and the atomic masses of their elements should furnish precise information about their characteristics. In this regard the distinction between atom and molecule was not as yet clear, and Berzelius spoilt an otherwise excellent system by refusing to recognize the polyatomicity of certain molecules. Nevertheless, the fundamental principle of a positive and generalized symbolism was at long last secured.

3.6. In this regard it is worth stressing an instructive distinction between written symbolic language and spoken nomenclature. It will be noted that parallel to the attempts to establish a graphic language, a nomenclature was developed. An attempt was made to formulate rules for the systematic construction of the names of compounds. Did this amount to a simple arrangement of ordinary language, or to the establishment of a scientific language superimposed on a written symbolism, which it would only translate? In fact, at the first stage of the symbolism, the parallelism is obvious. It is the qualitative composition of the compounds which must appear both in the name and in the graphic symbol. A certain superiority of oral language appears even in the structural indications already furnished by this nomenclature. Guyton de Morveau's paper on chemical denominations (Guyton 1782) was the first thorough attempt at a nomenclature of this sort; the formation of names of salts, in particular, is governed in such a way that their chemical nature shows through and the creation of new vocables, progressively with discoveries, is made easy and systematic.

A more extensive attempt in this direction was undertaken by Berzelius himself in his 'Essai sur la nomenclature chimique' (Berzelius 1811). For more flexibility and universality, he returned to Latin, and the endings ingeniously given to each stem took on very elaborate structural significance. The guiding principle was the degree of electro-positivity manifested by bodies in electrolysis. Thus, in the combination of two 'combustible' bodies such as sulphur and copper, the ending '*etum*' was given to the more electropositive, while the other was put into the genitive: thus, *sulfuretum cupri*.

The result is perhaps the most perfected form of the oral language of chemistry. If it was not retained, it was because the nomenclature of Guyton and Lavoisier,[4] being simpler, sufficed for the needs of a science henceforth provided with a graphic symbolism capable of expressing much more completely both structural properties, including those relating to quantities. The successive linear order imposed by spoken language is poorly suited to the representation of chemical structures; hence it will suffice to retain Guyton's nomenclature for designating substances easily and concisely; one resorts to graphical symbolism for a more complete scientific representation. Indeed, the progress of this symbolism soon rendered it completely irreplaceable.

3.7. The idea of valence, which was substituted for that of electro-positivity, led in fact to the utilization of all the possibilities of the graphic symbolism. It was not only the juxtaposition of the element signs which henceforth was significant, but also their positions in a two- or three-dimensional space. Auguste Kekulé, dozing on the upper deck of the Clapham Road bus, saw in a dream a dance of interconnected atoms, and the theory of atomic relations was born.

Already the simple system of juxtaposition permitted the distinction of substances of a roughly identical composition, by variously grouping their elements on one and the same line. Grey hydrated chromium chloride of Recoura was noted as $[Cr(H_2O)_6]Cl_3$, while the green chloride, of the same composition, which, however, loses two molecules of water through dehydration in a vacuum, is written: $[Cr(H_2O)_4Cl_2]\ Cl\ 2H_2O$. A third green chloride possesses an easily eliminable water molecule and is expressed: $[Cr(H_2O)_5Cl]Cl_2H_2O$. Simply by consideration of these structures, in which the number of molecules grouped around chromium in the complex ion is always six, the chemist is led *to assume the possibility* of a fourth, non-electrolytic formula: $[Cr(H_2O)_3Cl_3]3H_2O$. And experiment confirms this presumption (cf. Champetier 1943, p. 124). But such considerations could only have appeared after the discovery of the possibilities of structural expression of the graphic language of chemistry.[5] It will be recalled in passing what perspectives were opened up for chemists by the notion of isomer, which permitted the representation of distinct chemical and physical properties by the different arrangement of signs in two- or three-dimensional space, and thus enabled the chemist to predict certain of these properties by a systematic variation of the relative positions of the symbols.

Certainly, it would be dangerous to allow the belief that by these means the chemist can do without experiment, and reach a 'characteristic', the abstract study of which will deliver up all the secrets of nature. In reality, the success and the fertility of a scientific language always depends upon its ties with experience, and the history of chemical symbolism only confirms this necessity. From my point of view, I have simply wanted to emphasize the stages in the formation of a truly scientific language, which consists mainly in the conquest of *dimensional pluralism*, permitting the simple and fruitful expression of complex structures by means of which science explains experience.

From here on we can understand that ordinary language can serve only as a commentary and an accompaniment, a commentary moreover indispensable to the extent that it allows the linking of the conventional syntax of symbolism to experimental operations and results.

REVERSAL OF THE RELATIONS BETWEEN ORAL LANGUAGE AND WRITING

3.8. In the previous chapter I defined a linguistic expression as a *linear* sequence (2.15), which has, as a consequence, a single dimension spread over the course of time. Of course, I noted that the whole array of accompanying secondary signs – variations in tonality, in the intensity of the voice, mimicking, socially created affective associations within language – place spoken expression within a universe of several dimensions. But these new dimensions, which allow the variation of the concrete meaning of the same word, the same phrase, remain foreign to language considered as a regular vehicle of information. The different variations of tone which associate an affective nuance with a word do not constitute a *dictionary*, and they cannot be numbered in descriptive *units* of meaning. They offer a flexibility, a richness, an elasticity of expression which makes the use of the word an art, but at the expense of a certain fuzziness, a certain indeterminateness in language. A good part of these para-linguistic inflections are lost in writing: it does not recognize these *appoggiature*. But insofar as it constitutes only a simplified code of transcription, it in no way implies the neglect of these accompaniments. In the mind of the reader, writing no more suppresses them than the ancient semitic alphabet suppressed the vowels in the reading of Arab or Hebrew words. Ordinary language remains, even in writing, an instrument of information to which is added an inorganic ensemble of means of suggestion.

3.9. The same could not be true of a scientific language, whose function is precisely to preclude the introduction of arbitrary suggestion. Everything must be noted within it, all meaning must be governed by a defined system of semantic and syntactical rules. It is clear that a language fixed in graphic signs is in this case privileged. But it is no longer a question of considering the graphic as merely an auxiliary of transcription. Quite to the contrary, it is necessary to reverse the usual relations of language and writing, for scientific discourse is primarily graphic, and its oral form is somehow only a coded, approximate translation of the original.

Just as the written code of natural languages leaves something out of spoken expression, so the oral transcription of scientific language can only imperfectly note its content. Naturally, it is always possible to perfect a code and to make it adequate at the price of a multiplication of signs, at least to the extent that the content to be transmitted can be divided into discrete segments (which in fact is hardly the case in concrete oral expression, as we have just noted). At a pinch we can conceive of an exact oral codification of scientific language. But its original graphic form is the simplest and the most natural, because spatial extension permits the regulated and rigorous use of several dimensions of meaning.

While the anarchic multi-dimensionality of oral expression leads rather to a poetic overdetermination of vocables, the multi-dimensionality of the graphic signs of the chemist or mathematician evokes an enrichment of semantic and syntactic conventions. In the simplest cases it is certainly easy to translate the two-dimensional symbolism of mathematics into a linear sequence of oral language. 'x^2' is read without notable inconvenience as 'x squared'. But as soon as the syntactic content of symbols becomes complicated, reading only serves to designate very imperfectly in sounds the graphic symbol which constitutes the real language (just as written letters somehow serve to designate the phonemes of spoken language). What the mathematician writes as:

$$\begin{pmatrix} a_{11} & a_{12} \\ a_{21} & a_{22} \end{pmatrix}$$

can be described as "the squared matrix of two columns a_{ij}." But the operational schemata that characterize this type of mathematical object are spontaneously related to the two-dimensional sign rather than to the oral paraphrase. Now, the semiotic value of the dimensions of the symbol is even more obvious in the case of chemical language whose history has just been sketched. As a result there is a great temptation to draw metaphysical conclusions from this utilization of space.

MULTI-DIMENSIONALITY AND SPATIALITY OF SIGNS

3.10. It is well known that Bergson insisted vigorously on the importance of spatial reduction in the sciences, in contrast to the temporal immediacy of intuitive understanding. That scientific discourse has had open recourse to the multi-dimensionality of extension, seems to support Bergson's thesis. However, the question requires closer consideration. Is the spatiality which effectively exists in the symbol the same spatiality which is directly bound to our schemes of biological existence and to which Bergson refers? To the extent that linguistic expression is an action in the world of things and persons, the answer is certainly yes. But if we admit that this spatiality sets us free, provisionally, from the conditions of direct action, that it constitutes the most effective *detour* invented by man, we cannot help but see that all its spatiality consists simply in an abstract scheme of order and dimension. Employing these notions, mathematical reflection has made very great progress. It has shown their diverse meanings and has reconstructed them on the basis of only the simple non-intuitive notions of set theory. Yet it is not under this thematized form that a geometry of language need be envisaged. Just as a philosophy of mathematics must distinguish between the axiomatized notion of number as a theme of thought, and the systems of acts which govern its use as a scheme, it is also necessary to distinguish the thematized concepts on which the axiomatization of geometry is based and the naive schemes of order, discontinuity, dimension, which are at work in language. All language, natural or scientific, embodies a 'geometry' in this latter sense. Ordinary language assumes essentially the notion of a discrete and totally ordered set of signs; scientific language plays on the notion of 'dimension', the 'space' of meaning in which it deploys its multi-dimensional signs.

3.11. Under these circumstances it is apparent why science and its language are so closely and originally interdependent. The same naive 'geometry' which is for the one a condition of use is for the other the fundamental determination of an object. Whence derives the formalist interpretations of science, which view this relation from only one side, as if the linguistic aspect of our activity were primary, by comparison with things. No great leap is required to see the commonality of structure that prevails over both our activity of expression and our activity of determination and mastery over things. The formalist illusion stems from the desire to confer on themes, once they are disengaged by means of axiomatic abstraction, an ontological primacy over the operations which in fact engender them. One interpretation of scientific language and through it of science itself, as the simple reflection of a spatially predetermined structure of the object, appears to me to be no less erroneous. We must hold both ends of the chain, and understand that if the object is in fact given as transcendent object, all structure is disengaged in the course of an activity on two levels, first as a system of naive relations between manipulations or linguistic acts, second, as the objective theme of an elaborated understanding. There is an apparent vicious circle in all thought which requires foundation, for the activity of legitimizing thematized structures assumes schemes of the very structures which it is engaged in elucidating. Thus, the logician who demonstrates the non-contradiction of a theory must face the objection that his demonstration implies the naive use of certain schemes which are to be found thematized as objects of the theory itself. However, this is an unreasonable objection, for the notion of non-contradiction could only be precisely applied to themes, and not to operations whose guarantee is of another order, derived from an immediate relation with the object. Now, schemata adequate for an objective manipulation in no way imply, *a priori*, the coherence of the thematized constructions in a language abstracted from the manipulations. It is this formal coherence that the logician and epistemologist are concerned with establishing or refuting. The validity of schemes is a *condition* of linguistic activity, and the coherence of thematic structures is defined *by means of, and within*, a language.

Note that the difficulty that I am trying to express here involves two degrees. Logical thought can reflect upon itself, on the one hand at the level of an interpretation of science – and pre-eminently, of mathematics – because as a language, it can split into act and object; and on the other hand, at the very level of the linguistic act, because it distinguishes the effective conditions of language and λεκτόν [the spoken], the thematized language. It is the latter level that interests us, but it is important to grasp this analogy. In these conditions one cannot decide from the multi-dimensionality of language in favor of a subordination of the mode of expression of the sciences to the content of one of the sciences – geometry. For geometry itself can only be thematized as a science through the mediation of language.

3.12. Having articulated the nature of this 'spatiality' of the sign, it remains for us to examine briefly what innovations it provides for scientific language. To be sure, even natural language involves a 'geometry', as we have noted above, one which includes the schemes of the linear order. Scientific languages, that of chemistry as that mathematics, occasion the intervention of richer schemes, which are those of the partial order, governing the disposition of elements along several dimensions. But it is never a question of simply an *order*, a scheme relatively easy to thematize according to an abstract structure. The as yet uncertain signs introduced by the psychologist or the sociologist conceal no mysterious fruitfulness. The language of science is a written language, but the history of chemical symbolism has shown us what separates it from magical ideography.

Extension and duration, which appear to us only through their relation to qualified contents in our perceptive experience and in our immediate action, intervene here only as dimensions of expression, as the degrees of freedom of the elements of language. Of course, the arts of speech and writing play on this ambiguity, and literary discourse, written or uttered, acquires its virtue from this indetermination of its status. It deploys itself both in abstract space and time, which are the dimensions of language, and in duration and extension, which is filled by our movements and our passions. Whence follows the strange value accorded to speech, and the mythology of the sign. It is clear that our concept of scientific language should dismiss these phantoms, even while our sensibility takes pleasure in evoking them.

SEMANTIC POLYVALENCE

3.13. The multi-dimensional order of scientific discourse is one of its specific characteristics. It is important to bring to light a second which is no doubt very surprising, and even paradoxical. It consists, in different degrees, in a sort of semantic pluralism which seems at first sight the very negation of the function of expression, and more particularly of scientific expression. We shall explain this point.

Certainly, the language of science shuns ambiguity, and if one can speak of the plurality of meanings for the same symbolic expression, it should not be in the sense of an overdetermination of vocables, as happens in ordinary language. It is through the example of mathematics that this phenomenon is best understood. Since the beginning of the last century, mathematicians have delineated a general theory of algebraic operations, considered as laws of composition of any abstract elements. The prototype of such a theory is obviously ordinary algebra, whose fundamental operations are the 'four rules' of arithmetic, bearing on undetermined numbers. But it is precisely when one decides to apply analogous operative schemata to elements other than numbers that the modern notion of algebraic theory makes its decisive progress. Consider, for example, the idea of *permutation* of objects belonging to a set (a, b, c). Call P_1 the permutation which turns the order a, b, c into the order b, c, a; P_2 that which rearranges a, b, c into the order c, a, b; finally call I the permutation which leaves the order of the three objects invariant. (The identity permutation.) It is clear that the successive application of P_1 and P_2 to any initial order reproduces the same order:

$$P_1(abc) = bca$$
$$P_2(bca) = abc = I(abc).$$

Thus one is led to speak analogically of I as the 'product' of two permutations P_1 and P_2. A more detailed study makes it apparent that this law of composition has features generalizing those of arithmetical multiplication. Moreover, the same notion is applied to elements other than permutations or numbers, for example, by means of suitable definitions it is applied to displacements in space. Thus, a theory can be constituted with a precise symbolism, but one whose field of interpretation will be multiple. The notions of group or of ring can be interpreted as systems of operations on numbers, or on geometric shapes, or on matrices, etc . . . .

It is no doubt the possibility of these successive extensions, in which the *form* of the calculations remained the same, whereas the *nature* of the mathematical entities subjected to these calculations varied considerably, which was responsible for the gradual isolation of the guiding principle of modern mathematics, namely that mathematical entities in themselves are of little importance; what matters are their *relations* . . . (Bourbaki 1974, p. xxi).

It is in this way that the language of general algebra is used to describe structures, not entities, and in this sense that its semantics is ambiguous. It should be explicitly stated that this is even a matter of radical ambiguity. It is in fact conceivable that a language can be interpreted by means of several universes of objects, but that, in this case, although formed of different entities, the two universes are, so to speak, superimposable. That is, one can establish a one-to-one correspondence between their objects, preserving the property of being a result of an operation. (If x, y, and $z = x.y$ are in E and if x', y' and z' are their respective correspondents in E', then $z' = x'.y'$.) The dot '.' represents here the operation in E' or in E. The two structures are then called *isomorphic,*[6] and the semantic ambiguity is taken in a weak sense. But the language of modern mathematics admits a strong ambiguity, such that the same theory – that of groups, for example – can be interpreted in non-isomorphic and really distinct universes.

3.14. This polyvalence characteristic of modern mathematics is to be found to a lesser degree in the languages of diverse sciences. There is the famous example of Maxwell's development of an optical interpretation of the equations of electro-magnetism. In general, scientific expression disengages systematic structures which a tacit principle of semantic tolerance suggests extending to new domains. This is one of the reasons for the efficacy of scientific symbolism. Just as the ambiguity of the over-determination of ordinary language generates poetry, so the ambiguity of polyvalence of formalized language generates the restructuring of the scientific object. The process is at bottom more complex than the simple example of analogous structures might lead one to believe. It is a movement and reciprocal opposition of interpretations and of abstract structures expressed by symbolism rather than a simple comparison. Quite often, it is the mathematician with his highly formalized language who provides the structures; and the less abstract and only roughly sketched structures of the other sciences are then considered as possible interpretations. But this dialectic can only be established if the domain of objectivity of a science is already sufficiently developed in its specific language. We know that Aristotle denied the physicist the right to use mathematical methods; this was because Aristotle conceived the object of physics through the eyes of a biologist: his language emphasized only structurations of a classificatory type, of the act-agency type or of the ends-means type. Only optical and certain other phenomena (cf. *Anal. Post.* 78b. 32) are described in terms which permit the relation of their structures to those of the geometers. Thus, we see Aristotle explaining, in spite of his repugnance, the semi-circular form of the rainbow by introducing a hypothesis of purely mathematical reflection (*Meteorol.* 376a. 1). Furthermore, in the same text, in order to give an account of coloration, he refers to a biological phenomenology, by introducing the 'strength' or 'weakness' of rays, in relation now to the eye as a physiological apparatus, now to the conditions encountered by the ray itself in its line of propagation. Similarly, the recent attempts to axiomatize genetics (Woodger) consist mainly in giving its language a formal rigor, which is compatible with the distinct grasping of relations which would allow one to think of the object through a mathematical type of structure. In any case, it is a question of bringing out the possibility of polyvalence in scientific language, of disengaging it from the all too rigid semantic determination nevertheless required by the initial rectification of ordinary language, a language already too burdened with often contradictory implications. But doesn't stressing this polyvalence lead to stripping language of all real content? This is approximately the extreme neopositivist thesis, with its assertion that formalized language teaches us nothing of the world, that all coherent language is tautology, and that finally one can *say* nothing about nothing. Science would be instantaneously reduced to an empty discourse. This problematical consequence can still be avoided, however, by according to syntax a positive importance: The polyvalent languages of science do not have for their principal function the designation of objects but the articulation of syntactic relations.

3.15. Everything concurs to make syntax the principal significant element of scientific languages. When in ordinary languages syntactical organization does hardly more than reinforce the redundance of vocabulary in order to reduce errors of interpretation and facilitate the reception of messages, in scientific discourse it constitutes the essential part of the transmitted content. This fact is obvious in mathematics, where the practical significance of the symbols tends to reduce to their function itself, to such an extent that this science is sometimes considered, wrongly, as a simple language. The manifest predominance of syntax makes it reasonable, in this case, to attempt the most rigorous formalization, that is, detailed explications of the rules of symbolic language. To what extent is such a formalization possible and instructive with regard to the other sciences?

It is appropriate to insist first of all on the functional opposition of two kinds of language, informational and syntactic. The first is essentially oriented towards a designating function, and its unity of sense is thus the *noun*. But if naming has always appeared to be the fundamental act in poetry, the same would not be true for science. Thus, in the passage from *Alice in Wonderland*, used here for an epigraph, the roughly semantic demands of the duck, who wants a thing for each word – worm or frog – and the necessities of a language using 'empty' words .... are contrasted. For a syntactic language it can be said that the unity of sense is the *concept*, that is, a network of structured relations. The scientific role of syntax is to prepare the canvas on which this network is drawn. At its extreme pole, represented by the language of the formalized logic of propositions, it only draws the framework of tautologies which serve as a guide for every constitution of concepts, without adding any semantic information relative to a universe of objects.

The effort to formulate scientific languages, which, in its broadest sense, consists in rendering a syntax maximally explicit, thus leads essentially to this reduction of the semantic function, which in logico-mathematical language results in almost total elimination, with the semantic reference remaining no more than at the state of pure possibility, of degree zero, an empty place for content. But recognizing the decisive importance of this syntactic construction does not lead inexorably to the formalist illusion, according to which, so to speak, syntax would work all alone. Scientific activity *makes* a language, but science is not only a well-*made* language. The syntactical organization is only an aspect of the construction of concepts, which construction always presupposes an irreducible process of the manipulation of phenomena. The recent evolution of the formulations of the language of physics confirms this: it is not, strictly speaking, a system of objects whose structure the theorist delineates by formalizing his language, but systems of observation and intervention which he attempts to reduce to an axiomatic system.

Even in mathematics, where the situation is different, since there the only aim is possible structures in general, the syntactic organization has a negative side which generates a dialectical process. The fertility of a syntax is in fact limited, in the sense that the perfection of its formulation characterizes theories which are provisionally *complete*. Classical geometry, and in a certain way, the theory of the Aristotelian syllogism are good examples of this. They are thus inert branches of science, which only a critical renewal of foundations, by mobilizing an ancient theory at the heart of a more comprehensive structuring, by establishing itself, for example, at a higher level of abstraction, can revive. Such is the dialectical role of axiomatizations, which makes possible a recasting of the primitive concepts they brought to light. We shall shortly attempt to clarify this role in regard to the sciences of man. But it is important to add here another way of breaking down hardened syntaxes; this consists in a revision of the systems of focusing on the phenomenon. The behaviorist enterprise in psychology furnishes a significant example of this process. It remains interesting in both its positive contribution and in its failure, since the radical attempt to define the psychic phenomenon in terms other than those of internal experience partially goes astray, to the extent that it results in a purely semantical construction, a simple translation from an ancient language, but not at all the invention of a new syntax.

3.16. Thus, scientific language is characterized by the predominance of a syntax which expresses a structure of objects. Under these circumstances, if it is possible to specify the rules of this syntax so that they lend themselves to axiomatic employment, it therefore must be admitted that a sufficiently perfected machine could churn out a scientific discourse. And this, as we know, is true. This discourse becomes a calculus in the broadest sense of the word; that is, it becomes a concatenation of regulated combinations.

It suffices that the machine can effect operations step by step so to speak; read and compare the results of these operations, search its 'memory' for prior data, choose symbols according to a fixed rule. The essential difficulty is to know how to formulate instructions *for* the machine which govern these different processes, in other words, transcribe the contents of a metalanguage into the primary symbolism that is the very stuff of the calculus. But every complete scientific discourse can only be described and prescribed through a metalanguage that has two irreducible aspects: one, which can be transcribed for the machine, and into which the program is built; the other is apparently of another order and can be called a 'decisional' metalanguage. To it alone is scientific practice effectively wedded, for it describes the epistemological situations which are unforeseeable from the mere structure of the 'calculus', that is, from syntax in a strict sense. Everything that concerns this syntax can without doubt be described in the language itself, and thus inserted among the 'instructions', even though, let me say, choices and decisions are presented to it. But they are only decisions in a previously plotted framework. The machine, given two numbers, can choose the smaller, or decide upon one operation from among several possible ones, according to the results of the preceding one; it cannot, however, be ordered in advance to produce unforeseen structural relations, nor to integrate within the conduct of the calculus an extrinsic experiment, equally unexpected in its form and nature.[7] Furthermore the syntax of scientific language does not provide the means of determining the very categories according to which the object of such a science is constituted. Language is here only an indispensable tool in the service of the total activity of scientific *praxis*, a tool whose usefulness depends on the construction and domination of a syntax, but which is in no way itself the source and the material of knowledge.

We shall now see how language plays its role in a more general perspective of the constitution of forms, by examining the *découpage* of the phenomenon.

CHAPTER IV

THE *DÉCOUPAGE* OF PHENOMENA

THE MYTH AND THE CONCEPT

4.1. In each society, the events that man takes part in are found to be spontaneously molded by language. Indeed, the further we go back to forms of civilization less penetrated by science than our own, the more we see that almost *all* events are those in which man participates, or more exactly, they involve an enlarged or transposed idea of human powers. This is the most certain meaning one can give to the 'primitive mentality' in the analyses of Lévy-Bruhl and the more recent ethnographic descriptions. The transition from events to *facts* is effected by the intermediary of the spontaneous use of a language, which is itself a very elaborate result of social life. The facts to which a nascent science refers as its data, the 'proto-scientific' facts, thus cannot be considered as presented directly in the network of principles and schemes of a transcendental subjectivity; and if it is necessary to search in science for the monogram of such a subjectivity, this search must be carried out at the other extreme of the process of knowledge: the transcendental architecture of the scientific object is the conquest of hard-working, trained thought. The original fact, on the other hand, is subject to the extrinsic determinations of a concretely lived culture, of a practice the objectivity of which is completely external and in no way intentional. The untrained thought of a man whose oxen pull a cart moves from the event that he lives to the *fact* of cartage, from which he extracts a notion of force, apparently associated with uniform displacement; practice, in the complex and confused conditions which determine it, justifies perfectly this primitive notion of force and effort. But to the extent that this practice is diversified and extended, to the extent that a more refined wish to analyze is simultaneously awakened, thinking schools itself in the conditions of the phenomenon and objectivity is internalized. The immediate transposition of the primitive notion of force into domains beyond that of the fact where the notion was found objectively determined is called into question. For this uncontrolled transfer of a synthetic notion to other situations viewed globally as analogous characterizes myth. The mythical notion incautiously enriches the fact with the overdetermination and the overabundance of meaning of a language left to itself. The dissolution of myth, to the advantage of the concept, consists then in a dissociation from the conditions of presentation of the fact, which then leads to a new *découpage*. The objectivity of a notion is no longer founded in the immediacy of the experienced event, but in a more and more conscious and more and more refined process, in which language and practice are controlled and dominated. The phenomenology of the scientific object cannot thus be an immediate and stable given, which it is only necessary to bring to light. It is presented as a mutating process of thought.

It need not be concluded from this, however, that epistemology is reduced to a genetic psychology. Epistemology is concerned with systems with objective aims, with phenomenologies in action; the genetic psychology of intelligence describes and analyses the phenomena of maturation, growth, and equilibrium which condition the transition from one phenomenology to another. It does not at all explain the contents of various phenomenologies but looks for the laws of the dynamic or, more modestly, of the kinetics of an intelligence engaged in a concrete biological and social experience. We will not concern ourselves now with this, so to speak, macroscopic origin of scientific concepts, but only with the internal organization of systems within a phenomenology. From this point of view, the *découpage* of the constitutive facts of a scientific domain appears to be, first of all, a legacy implicit in language. As a naive instrument of this *découpage* of facts, language naturally transmits myths, in the sense indicated above, as well as ideologies, i.e. more or less explicit justificatory interpretations of situations actually realized by social organization and practice at a given moment of history. Bachelard has clearly shown in his *Formation de l'esprit scientifique* this overdetermination of notions, of which thought must divest itself in order to reach the concepts.

4.2. In the domain of the sciences of man, we can easily see that this diffuse orientation imprinted by language takes on full force, for better and for worse. One can be tempted to reduce the observation of the human fact to an analysis of concrete language, under the pretext that language is the repository of all human reality. This is scientifically pernicious, and the brilliant speculations of the author of *Propos* are an example of this pernicious attempt: it is one thing to offer a philosophical account of the human condition, perceived through the traits of language, and quite another thing to construct an object of controlled observation and of experiment.

The *découpage* of human facts presents a special difficulty. Here the phenomena have an immediate *sense*, which means that they spontaneously take part in a universe of valorized and directed actions, either in the consciousness of an individual, or in the organization and functioning of a collectivity which is given as a whole, even when the relations of this whole escape us. This sense is transmitted by language for the speaking subject of each social group, and it is this that constitutes, for our consciousnesses as agents, the very essence of the given human fact. In this way the proto-scientific human fact is already presented as provided with a structure, and as a pseudo-object of science. An explanation of these meanings, generally retouched in order to give them the coherence which they lack, can provide the illusion of a scientific understanding. The organization of lived human *experience* [*vécu*] by means of both meditation on these meanings carved out according to social practice, and essentially by language, offers itself fallaciously as the object of science. A completely analogous epistemological situation would present itself – has been presented – in the natural sciences, if the object of physics were defined as a complex of qualitatively experienced sensations, that is, if a phenomenology of the perception[1] of things were substituted for the science of objects. But if the culture of our time makes us apt to challenge this paralogism immediately, it has not yet sufficiently prepared us to perceive the analogous illusion in the domain of human facts.

The purpose of this chapter is to discern the actual paths through which the resumption of a scientific *découpage* of these phenomena is effected.

EXPERIENCED MEANINGS AND SCIENTIFIC OBJECTS

4.3. It is useful to begin with a brief general examination of an example of this epistemological resistance of the human fact. Let us borrow an example from Lévi-Strauss (1945). In the treatment of the notion of *avunculate*, ethnology furnishes us with a good example of this dialectic of signification and phenomenon which governs the *découpage* of scientific fact.

In numerous societies, the maternal uncle plays a very special role, in the sense that his relations with his nephew have a definite character. The ethnologist is thus spontaneously led to postulate a parallelism and a rigorous correlation between the system of kinship terms used in a society, and the system of attitudes through which relations between individuals or classes of individuals distinguished as such are expressed: the human fact is carved out here according to the immediate indications of social practice deposited in language.

But when the ethnologist searches for an explanation of this privileged character of the maternal uncle/nephew relation, he turns towards mechanistic hypotheses, justifying the present fact by the residual action of an abolished social structure. For example, he interprets the avunculate as a left-over from a matrilineal regime, or as the residue of a marriage between cross-cousins; now such hypotheses are either very doubtful or are inadequate for giving an account of these observations . . . Lévi-Strauss, pursuing some remarks of Radcliffe-Brown, concluded that the very notion of avunculate did not adequately account for the fact to be studied. There is a correlation between the attitude of the nephew towards the uncle, and that of the son towards the father; moreover, the same is true for two other types of attitudes: brother-sister and husband-wife, so that the system of these four organically bound pairs constitutes the scientific fact to be studied as a whole. Lévi-Strauss believed himself able to enunciate a law of compensation among these four relations, of which two would always be free and familiar, while the other two would be more or less hostile and antagonistic. The avunculate would be only a partial aspect of the structure of equilibrium which is instituted in the 'atom of kinship' which necessarily includes the man who gives his sister, the man who receives her and the son of the couple. My aim is not to appraise the value of this interpretation so given, with respect to its content, but to restrain the strange approach which leads to a restoration of a scientifically describable human fact on the basis of a raw datum cut out by ordinary language. In terms of the ethnological analysis, one rediscovers a fact provided with meaning: but it is not the same meaning transmitted directly by language and experienced in social practice. A new system, a new phenomenology of the scientific object is substituted for the crude system of experienced meanings, for the phenomenology of perceived relations. The radical difficulty for the sciences of man derives precisely from this necessity that the scientist's goal is facts provided with sense, but he can only attain these facts by developing data which are already meanings at the level of immediate perception. The double temptation that awaits the scientist is either to remain simply at the level of events as experienced, or, in an inappropriate attempt to attain the positivity of the natural sciences, to liquidate all meaning, in order to reduce the human fact to the model of physical phenomena. The constitutive problem of the sciences of man can be described from here on as the transmutation of experienced meanings into a universe of objectivized meanings. From this fact is derived the fundamental importance, beyond their gropings and inadequacies, of new disciplines like information theory and cybernetics.

ORGANIZED PRACTICE, THE CULTURAL ENVIRONMENT OF THE CONCEPT

4.4. Such a characterization of the problem brings out clearly the determining importance of relations between theory and practice. As long as the *découpage* of human facts remains a tributary of the language and the naive ideologies that it subtends, knowledge can only describe the apparently spontaneous adaptation of behavior, in a practice where the agent does not clearly distinguish an object from an image produced by the diffuse ideological interpretation of his own activity. Ancient 'psychology' and 'sociology', with the exception of admirable passages in Aristotle, remain almost consistently at this level. The dominant preoccupation is that of a *political* organization of human life. For these first attempts at a science did not remain fundamentally speculative as did later attempts. But the notion of *city, citizen, economy* are here directly determined by the play of social and natural forces which proto-scientific thought reflects and justifies without analyzing. The human fact is grasped in its immediate appearance, as the generalization of an experienced event. The knowledge which results thus oscillates between two poles: that of artisan-like technique wedded directly to the phenomenon, without rising to the concept which objectifies it and discloses its sources, and that of a wisdom stemming from meditation on meanings, which tends to blossom into a utopia. The first form of knowledge is of the same order as that of the artisan, who *lives* the tricks of his trade, and carves up the universe of tools and processes in which he executes them, in conformity with the mythic kernels of meanings borrowed precisely from the experience of human relationships. Naturally the larger part of our knowledge of the social world and our own reactions, even in the heart of a culture so penetrated by scientific thought as ours, remains of this type. It is necessarily the daily bread of our individual practice, and just as our advanced sciences of natural phenomena transform our concrete grasp of the physical world only partially, imperfectly, and progressively, so a science of man, once organically constituted and collectively employed, could not prevent us from living a good part of our lives according to immediate meanings and myths. The scientific universe, it is true, still penetrates our lives, but not as such, for it is barely imposed only as it renews rather than as it suppresses these meanings and these myths. It seems that scientific knowledge, the effective concept, instrument of controlled practice, becomes assimilable into individual life only by changing itself until it takes on the very qualities of naive notions. This process is undoubtedly unavoidable, but dangerous for the human sciences, certain of which run a continual risk of falling back into their primitive state. A concept like that of *class*, for example, whose scientific elaboration is far from complete, returns spontaneously to myth in its daily usage. However, as the result of a new *découpage* of sociological data, and proceeding originally from a critique of the direct perception of social facts through the vehicle of language, the scientific status of 'class' can be preserved only at the cost of a constant effort of analysis and structuring of the facts. Its very success, in social practice, at the technical level of political struggle and ideological polemic, causes the notion of class to penetrate into ordinary language which swallows it up in the hidden set of meanings that language transmits. The concept, barely constituted, insufficiently bound into a structure of the scientific object, returns to the diffuse and confused 'wisdom' of language, which is a sort of Madame Tussaud's of real knowledge.

At the level of practice, the scientific concept seems thus to be necessarily adulterated, and from this adulteration no doubt stems the traditional opposition between a pure science and a practice. But to the degree that science develops, another level of practice is organized, on a collective scale. It is this practice, planned, structured, organic, which is the true cultural environment of the scientific concept; within its scope the concept is definitively constituted, developed, refined, transformed; for the perception of the experienced datum is substituted the grasp of an objective datum, for which the *fact* is a controlled product. In the domain of the physical sciences, this institution of a coherent practice is expressed socially by the transition to industry, whose work is oriented not by mystically experienced notions but by systems of objects carved out by learned consciousness. For the sciences of man, this transition is certainly more delicate, and only its premonitory symptoms are as yet discernable, but it is indeed this transition which should permit the future rise of a scientific understanding.

4.5. Following upon the naive *découpage* of human facts according to spontaneous practice and linguistic tradition, repulsion in regard to all practice seems to manifest itself at once. This is the radically idealistic moment in the constitution of the object, which is often expressed in the present context by a bias towards mathematiziation in the extreme. Certain aspects of the Platonic conception of the social fact clearly depend on this, although the contemporary state of mathematics has made such an enterprise too crude. The social phenomenon is then described, sometimes on the mode of experienced relations, and in particular the relations of authority at the heart of the domestic and political group, sometimes in the perspective of a semimystical arithmetic, in which the style of Platonic expression hardly permits a certain distinction between analogous image and foundation. One then rediscovers, in the course of history, this indistinct combination of a phenomenology constituted at the level of immediate experience and naive ideology, and a resolutely abstract phenomenology, focusing on objects of a mathematical type. Economic notions offer perhaps the most striking examples of this. For the mercantilists the economic fact was essentially viewed as an exchange of a commodity for hard cash, and it was the practice of the merchant that dominated the analysis. But on this immediate meaning of an experienced activity was superimposed the rough numerical schema of the balance of entries and withdrawals of money. The determination of economic laws boils down to the study of the conditions and factors of a cash surplus for the national group. The effort to arithmetize mercantilist economics is at bottom correlative to an attempt at the political direction of exchanges; but the integration of the two orientations is ineffective. Now, the definition of the object of a science of man requires precisely this integration, which constitutes a more advanced moment of its progress. Abstraction then is no longer exercised on things supposedly isolated from human action, but on the schemes of action themselves, which become the *themes* of an objective analysis. But science ceases to be a simple transposition of immediate practice into mythical notions; it aims at practice, certainly, but at an objectified practice to which controllable and perfectible instruments of analysis are applied. This thematization of schemata of action, viewed first of all as experienced on the mode of Bergsonian intuition, on the one hand certainly shocks common sense, and on the other, requires for its effectiveness an effort to invent new abstract structures. Oriented towards the construction of models for physical phenomena, human mathematical genius turns towards its new task uneasily. Nevertheless, everything seems to indicate that it will succeed.

AN EXAMPLE OF STRUCTURAL OBJECTIVATION: THE 'WAGER'

4.6. We shall try to show, by a brief example, the direction this profound transformation of the phenomenology of the human fact. The action of *betting*, under the diverse forms it can assume, is certainly universal among men, but before the seventeenth century it does not seem to have been systematically carved out as a fragment of behavior, as the object of a knowledge of the human fact. As we know, it was principally in the work of Pascal,

Fermat and later Jacques Bernoulli that this action was raised to the level of object, and passed beyond the universe of experienced behavior to the universe of thematized structures of analytic thought. Nevertheless, this transmutation still remained sufficiently ambiguous for the notion to be used in the famous Pascalian fragment so as to reveal the oscillation of thought between the *wager*, as an experienced aspect of naive practice, and the *wager*, as a mathematized concept. Pascal analyzed a scheme of conduct whose field of application is the world of immanent human experience, and he extended it to the transcendent domain of a religious experience. Of course, the extension is presented only as a subsidiary argument of apologetics, aimed at showing that rational behavior is not at all incompatible with faith, rather than that it favors or disproves it. This is accomplished by means of a particular insistence on the existential side of the bet, on its aspect as an experienced moment. Pascal's originality lies in having wanted to juxtapose this perspective of "we are on our way" with a structuring of the conduct of the wager that involves an evaluation of the stakes and the chances, and a calculation. But the transcendent character of the domain of application renders this determination illusory and the schema is out of true. Nevertheless, there remains the idea of this analysis of behavior, placed in a situation of immanence, which appears to us today as an admirable prototype of conceptualization in the sciences of man. The experienced notion of decision in an uncertain situation becomes the object of thought, a structured theme which is dominated by a mathematical analysis yet to be advanced, and of which the genius of Pascal invented a first form.[2] Let us therefore consider Pascal's wager from this point of view and we shall see the obstacles which stand in the way of the attainment of the concept.

The elementary Pascalian model of decision making is apparently that of a lottery. The subject chooses from among several 'tickets', knowing each of their 'chances' of winning, and the value of the win. Thus, it is sssumed that the probabilities inherent in certain events are *objectified* and determined, depending on a mechanism external to the subject. In the case of the transcendent wager of the *Pensées*, this schematization obviously faces the insurmountable difficulty (at least from the perspective of the non-believer) of assigning an *a priori* probability to events: "God exists" and "God doesn't exist". In the course of developing his argument Pascal successively attributes to these events different hypothetical values; but these values do not intervene in any essential way in the calculation of the decision. This is why the model of the lottery imposes an inappropriate schematization here. Certainly, in regard to the wager experienced by the player vaguely postulating a certain predetermination of the outcome, which a mysterious grace might somehow allow to befall him, the objectivation of probablities which would be materialized by the distribution of a collection of balls in an urn clearly constitutes a first and quite positive rationalization. But in the more general and more natural case of a subject making up his mind under conditions of uncertainty, the objectivation of probabilities in an external system becomes myth, produced by an overhasty transposition. The analysis of the situation, relieved of its existential content and freed from the influence of a model then socially predominant, calls for a new schematization, one which is brought to light by the game theory of strategy. The Pascalian wagerer has no right to postulate *a priori* the chances of gain and loss, in order to make the choice which maximizes his 'hope' of gain.[3] He can only set up a table of eventualities which he foresees, according to the very nature of the trial and the tactics which he can follow; the combination of possible events with the consequences of his own choices presents him with determinate results of gains and losses. Rational action will consist here in assuring, for all possible outcomes, the greatest possible gain and the smallest loss; it is thus wise to assume that the independent outcome, governed by an evil spirit, is determined in a way which places the better in the most unfavorable position, and to make a decision by drawing the best part of this situation. Such is the fundamental principle which governs the models of rational decision making in game theory.[4]

A mathematical calculus derives from this, which reintroduces in a singular way the notion of probability, since it demonstrates that in general the optimal choice among various strategies must be aleatory choices, attributing to each of the strategies chances indirectly determined by the conditions of the game.

Thus one finds a human fact transported from the plane of lived experience to the plane of objective themes, which, from now on, belong to science. All the objections which this epistemological transmutation can meet are evident; the mathematized scheme retains none of that affective warmth that the event had in daily life. It reduces the fact to only one of its aspects, i.e., the rationalization of behavior: now almost no effective action has as its sole significance the search for an optimum. These are objections which have some force, but which are not pertinent. For the ideal of science such as we can formulate it today is not to *substitute* for the experience of a consciousness the necessarily abstract objects which this consciousness sifts out; it is sufficient that these schematizations permit the description of controllable interconnections at a certain level of experience. It is true that from one point of view, scientific models of facts are partial, but they are essentially perfectible, and if they are valid, they include within themselves the precise indexes of their limitation, their degree of approximation, their lacunae.

The *découpage* of the fact in the sciences of man thus involves a spectacular metamorphosis of the perceived datum. It is, however, at this price that science is possible, even if it were at the humblest level of the description and classification of phenomena.

TWO APPARENTLY OPPOSED MOVEMENTS: 'FORMALIST' *DÉCOUPAGE* AND 'OPERATIONAL' *DÉCOUPAGE*

4.7. It appears then that the problem of *definition* in the sciences of man, even more than in the other empirical sciences, constitutes an essential stage in their progress. It can only be resolved by the advent of a scientific *practice* which is resolutely detached from the immediate practice informed by language. The specific difficulty of the sciences of man is certainly connected [to borrow a concept from gestalt psychology] to the self-imposed force and stability [*pregnance*] of the experienced forms of the human fact in a given civilization where the phenomenon is spontaneously grasped as *myth* in the sense of the preceding analysis. The effort of contemporary scientific thought to establish a knowledge of man can be described initially as this attempt to transmutate the ordinary focus on the facts, resulting in an often original *découpage*, which would finally place the phenomenon at the level of scientific objectivity. At present, this effort seems to be deployed in two still distinct but probably convergent directions; the preceding pages have been intended to prepare us to grasp their significance. On the one hand there is clearly a movement towards a formalistic *découpage* of the phenomenon, reducing the object to abstract structures, which are radically separated from the experienced or perceived phase, and thus they leave themselves open to criticism from the champions of the concrete, of the living, of real existence. This formalist *découpage* of facts apparently intends to attain in a single stroke the type of construction which a long series of trial-and-error, revolutions and successes established in the natural sciences. The other movement, still rather new, can be christened the 'operational' *découpage*, from the term given it by one of its most significant aspects. It consists in attempting a relatively concrete but limited synthesis of the construction of a formal model and practice. The operational method was born of the encounter between rational thought and the most down-to-earth problems posed by the organization of human work. As the optimal utilization of machines by man, organization of communication in a group of collaborators, distribution of the means of action in the face of an uncertain outcome, 'operations research' provides an effective way of thinking about human facts. It is a mode of access which can serve as a paradigm for an enterprise much less pragmatic than operations research has been in its initial stages, and can open new perspectives to certain parts of science. By examining the two opposite poles of formalism and operations research from the perspective of their determination of the human fact, I shall try to outline the movement which in fact brings them together towards an epistemologically more adequate *découpage* of the phenomenon. In both cases formal thought is present, and in both cases it plays the role not of an ideal of knowledge, nor of a final cause of scientific research, but of a dialectical instrument of provisional opposition to 'the given,' at whatever level it may be; it plays the role of motor-cause of knowledge.

4.8. I have chosen as a first theme of this epistemological analysis one of the apparently most formalized enterprises of the current phase of science: structural linguistics. The attempt to transform the experienced event into an abstract object, defined essentially by its correlations with other objects in a formal system, seems here to have been pushed to extremes, and is presented as a veritable provocation in the face of the customs of empirical knowledge. Thus it is imperative to show its implications, its difficulties, and its true meaning. This will lead us to bringing out the dialectical character of its form beneath its apparent rigidity, and this dialectical character manifests itself in two principal aspects:

First, in the opposition between different levels of the formal, an opposition which relativizes them and preserves the notion of form from a Platonic interpretation. The use of forms in science could only sustain an idealism through the effect of an ideology external to the progress of knowledge.

Second, as the development of a diachronic theory, of a dynamic – or at least of a 'comparative statics' – of systems. The concept of structural equilibrium is elaborate, and furnishes the means for an initial interpretation of linguistic changes, considered in their intrinsic conditioning. Here is the point of junction between a formalizing theory and a more comprehensive discipline which would integrate the event into a structural universe. Attaining this junction is a major difficulty of the sciences of man but it can be clearly diagnosed only if one accepts as an essential and positive epistemological given the movement of formalization. It is to a description of this movement that we now turn.

THE SAUSSURIAN REDUCTION

4.9. As we know, the historical point of departure of the formalist *découpage* in linguistics is the Saussurian conception of *language*. The discovery, in the nineteenth century, of series of relationships on the one hand, and on the other, of systematic phonetic correspondences between languages supposedly of the same origin, no doubt opened up a new view of the object constituted by the fact of language. But this fact of language, if it is separated from the multiple overdeterminations that attach themselves to the experienced event of the speaking and listening subject, remains nevertheless awkward in its meaning: the fact of language appears to the linguist as essentially integrated in a longitudinal context like a historical fact, but its transverse *découpage* within an organically constituted system is carried out on the plane of immediate knowledge. Words, or sounds, elements directly perceived by the consciousness of the user, are identified as a function of philosophical traditions, habits of perception, and ideologies. Our purpose here is not to search history for the traces of systematic attempts at a scientific *découpage*, which transposes the elements of language onto the level of scientific knowledge; such traces certainly exist, and the example of the Hindu grammarians would no doubt be particularly instructive for studying a long-term origin of this metamorphosis. In his decision to define "language [as] a system whose parts can and must all be considered in their synchronic solidarity" (Saussure 1974b, p. 87), it was Saussure who openly posed the problem of a transverse *découpage*. The first objective of science henceforth was to be that of discerning the true elements of a language, and of describing their systematic relations. In conformity with the most general epistemological law, the scientific observation of facts is expressed in terms of its integration in a system: the authentic object of knowledge, its center of gravity, is the system. The Saussurian conception of language thus brings about a double reduction of language: reduction in relation to history (synchrony moves to the foreground of linguistic study) and reduction in relation to the psycho-social context (language is considered "in itself and for itself"). Such a reduction is as radical as the Galilean-Newtonian reduction of the physical fact to spaces, times and masses. However, it appears still more Draconian, since it eliminates from the linguistic object not only a complete set of probably determining conditions (as Galileo did for general mechanics), but also a mass of meanings perceived, in immediate experience, as essential to the use of words. But the scientific object can be constructed only at the price of this inversion of common sense. Prior to being able to treat language as a concrete historical event in a context of social relations, Saussurian linguistics proposes, as its object, language, a synchronic system of elements whose function is certainly significant, but which are studied first of all as parts of a system within which they are interconnected. The formal laws of these relations define a linguistic structure.

It would not be accurate to characterize this preliminary step as a simple *abstraction*. The linguistic structure focused on here is not only abstract in terms of the fact of language; it is what, for lack of a better word, one can call, following Husserl, an essence; that is, leaving all ontology aside, it is a transcendental outline of the object. Transcendental does not retain here any strictly idealist meaning, to the extent that there is no question of bringing to light an immutable condition of understanding of the object rooted in the nature of an abstract '*I*'. The transcendental outline is the foundation for the scientific understanding of the object, but far from fixing its form definitively, it never constitutes more than a provisional determination; in other words, it is involved in a 'natural' process of evolution of the understanding, which will eventually make of this an out-of-date or more or less incomplete goal. The word 'transcendental' is, however, justified from our point of view, precisely because the outline in question cannot be reduced to an impoverishment by means of an abstraction derived from experience. Whatever its genetic status, once established it constitutes the guide to conceptual knowledge, by *making possible* the contributions of a controlled experience and the development of a regimentation of these experiences. It certainly excludes from present-day science a horizon reached only by immediate knowledge, but once this negativity is recognized and as it is gradually made clear, it will engender new requirements that will lead to a mutation of the phenomenology thus adopted. A new transcendental outline of the object will arise from this, which will compete with the first, and in the best of circumstances, come to include it. The Saussurian view is thus a stage in the definition of the linguistic object, a starting point insofar as it constitutes for the first time a coherent transcendental view, a radical reduction.

THE PHONOLOGICAL *DÉCOUPAGE*

4.10. Since the historical phonetics of the nineteenth century offered the earliest examples of sufficiently rigorous linguistic regularities, it was natural that the Saussurian revolution was first concerned with sounds. Classical phonetics in fact took as its object the sounds of a language, either as it defined them physiologically through their conditions of production, or as it described physically their acoustical qualities. In both cases the linguistic object was considered as an isolated fact, and was described in physiological or physical frameworks that are not proper to it. At most a classification was introduced which distributed sounds according to their common characteristics. Historical phonetics would examine, for example, the transition from a class of unvoiced consonants to the class of corresponding voiced ones. This somehow involves *qualitative* determinations of the object, determinations taken in and for themselves in their positive aspects.[5] Paradoxically, the Saussurian revolution did consist in defining the linguistic object at the level of sounds, no longer in terms of their voiced qualities, but rather in terms of their mutual oppositions, which determine not isolated sounds, but a system where each element has a value only through its relation to those to which it is opposed. The object is the phoneme, an "oppositive, relative, negative entity" according to Saussure's famous definition. It is only under these conditions and according to this view that the sound is a linguistic object. Much later Bloomfield said "the importance of a phoneme, then, lies not in the actual configuration of its sound-waves, but merely in the difference between this configuration and the configurations of all the other phonemes of the same language" (Bloomfield 1933, p. 128). The phonological object is a structured class; that is, a relation between sounds. The members are not grouped together by the simple possession of a common characteristic: with regard to other sounds, they must still maintain analogous relations of opposition, making them structurally equivalent in the system. The phonological description of a language is thus not reducible to the juxtaposition of descriptions of its sounds as they are realized by speaking subjects. Here the scientific object is the entire structure. The experimental confirmations of the real importance of this object, which so easily escapes immediate perception, would not be lacking. These confirmations appear principally when one examines the behavior of a subject speaking a language other than his own, and consequently adapting himself to a foreign phonological system. If one sound exists as such in two languages and is the realization of different phonological functions in each of them, it can present the foreigner with unexpected difficulties of identification and reproduction. Troubetzkoi cites the example of the posterior open and rounded vowel (ɔ) in Bulgarian; it exists in Russian, but with a completely different function in the vocal system, and presents a considerable difficulty to the Russian subject speaking Bulgarian. Reciprocally, Roman Jakobson notes that as a Slovak and a Russian do not possess the the anterior closed rounded vowel that one hears in the French *jeu*, they perceive and reproduce it in conformity with their respective phonological systems: the Slovak, like a closed [é], because his system has only the grave/acute opposition, the Russian, like an [o], because his system rests on the rounded/unrounded opposition.

Phonological analysis thus results in resolving the linguistic object into a small number of mutually opposed distinctive traits, diversely grouped in phonemes, of which they represent, so to speak, the significant *dimensions*. The peculiarities of realization of these phonemes, which are not pertinent in the system, are either determined by the environment – but bear only redundant information – or they remain in this first approximation, relatively free for the speaking subject, and in no way condition the perception of the phoneme. The constituent body of language as scientific object is thus reduced to the oppositional structure of phonemes, a structure that makes possible an abstract arrangement, capable of bringing to light linguistic laws. The radically formalist stage is reached when it is stated that "it is therefore more convenient to consider the elements [of language] as purely logical symbols, upon which various operations of mathematical logic can be performed" (Harris 1951, p. 18).

4.11. There are, however, difficulties inherent in this point of view; and considering them brings to light its true nature. In the first place, the phonological reduction is achieved on the basis of a datum already elaborated by a prior phenomenology, not to be confused with that of the simple perception of sound. An initial *découpage* of the acoustical facts precedes it, which determines as linguistic proto-object either the sound, as produced by vocal articulations, or the sound as a vibratory phenomenon. Either perspective can serve as the point of departure for the phonological analysis. The first leads, for example, to distinguishing an opposition between voiced consonants and unvoiced ones, according to whether or not there is vibration of the vocal cords; the second, an opposition between *compact* and *diffuse* phonemes, according to whether or not the central periodic component predominates in the spectrogram. These distinctions are clearly heterogeneous. R. Jakobson adopted the second description of phonological traits, but many linguists are satisfied with the first, and the two are only imperfectly matched. Certainly the phonological point of view completely rejects these determinations in the proto-linguistic limbo, and as one phonologist forcefully claimed,

Every brute phonetic (i.e., non-phonemic) alphabet is a more or less traditional collection of signs so used as to reflect a few of the positional variants of some of the phonemes (Swadesh 1954, p. 85).

The problem nevertheless remains of choosing a unified phenomenology, and its current absence underlines the pluralistic character, or more exactly, the stratified character, of the *découpage* of phenomena.

Whatever may be the point of view adopted to describe the phonetic substructure of segmentation into phonemes, the logical development of the Saussurian conception leads paradoxically to presenting as the ideal limiting condition of the analysis of a language its unintelligibility for the observer. In a phonetic material deprived of its immediate meaning for him, the linguist searches for pertinent oppositions, that is, oppositions used as marks to transmit information, but without in any way grasping the content of this information as such. It is thus as a purely informational schema that he describes language, and this is the first object of a linguistic science. Certainly techniques must be developed which will permit one to decide experimentally, so to speak, on the organization of phonetic traits into systems of opposition, techniques which amount to comparative variations of the environment of a supposed phoneme. This is the meaning of Harris's bold attempt at an empirical description of a language by bringing to light its phonetic structure, independently of any hint of semantics. A sketch of his objectivation is as follows: tape-record raw material, for example, fragments of a spoken sequence. One ignores the meaning content of this complex of phonetic traits, but one hypothesizes that it is constituted by a series of phonemes defined in a network of oppositions; the phonology makes explicit the general conditions of realization of these oppositions, and tends naturally to organize them into a body of propositions which constitutes an axiomatic definition of the phoneme. It is in this perspective that the observer then analyzes the sounds of the examples of language and succeeds in describing its phonological system, that is, in constructing the linguistic *object* that he proposes to study.

It is apparent that, in its limited form, the process of objectivation consists in bracketing the semantic content with its immediate value in psycho-social experience, in order to reveal signifying structure and functions. Far from eliminating meanings, they are objectively recognized as the *raison d'être* of language, and it is their *conditions of possibility* which become the theme of a science of linguistics. Yet, the conditions of the possibility of the signifying function do not appear only at this level of the problem. The immediate experience of language is realized confusedly and syncretically at different levels, which scientific analysis dissociates and organizes. Taking account of this cleavage in different structural planes is essential for understanding the formation of the scientific object.

HIERARCHY OF PHONOLOGICAL STRUCTURES

4.12. At the very level of the phonological object, a dialectic of organization on several planes is sketched. The 'paradigmatic' unity of the phoneme is opposed to the 'syntagmatic' unities of higher order. A hierarchization of phonological structures would require the successive study of a phonemic system, a syllabic system, a system of 'words', as the units of higher orders are defined, like the phonemes, by the conjunction of relational traits. For example, a central and a non-central element are opposed, distinguished by a 'culminative' trait of some kind or other. We can thus consider the groupings of phonemes into syllables, syllables into 'roots', 'roots' into 'words'. The Spanish *miro* (I look), and *miró* (he looked) will oppose each other structurally in the following manner:

	miro	*miró*
non-central syllable	zero	/m/ /i/
central syllable	/m/ /i/	/r/ /o/
non-central syllable	/r/ /o/	zero

According to the Argentine linguist from whom I have borrowed this example (Prieto 1954), it is necessary to introduce two types of structural relation, characterizing the phonemic-paradigmatic level on the one hand, and the superior syntagmatic levels on the other. The first, as we have seen, concerns articulated *oppositions*, defined by a conjunction of traits. But superimposed on these oppositions between phonemes is the *contrast* between the central element, distinguished by a culminative trait, and the non-central element. The oppositional structuring thus would come into play only among elements having the same syntagmatic function; in a syllabic language, for example, the oppositional content of a vowel is defined only in terms of other possible central elements of a syllable; a phoneme that is incapable of becoming a culminative element, should its phonetic realization be very similar to that of a vocalic phoneme, is defined by a very different oppositional contention. The syntagmatic point of view thus involves a reshaping of the phonological object. The structuring of the first level is thus integrated and dominated,[6] so that its radically abstract and artificial character is henceforth placed in a dialectic perspective. Thus it becomes clear that on the phonological plane the analyst recovers the factor of meaning which he had to set aside ruthlessly at the outset; for the superior syntagmatic unit is the phonological 'word', which must be defined by reference to a content. According to Luis Prieto, the name 'phonological word' is given "to these syntagmatic units which compose the phrase on the level of expression, and which correspond to the units on the level of content which also compose the phrase" (Prieto 1954, p. 50). But this reference to content in no way constitutes the definitive point of arrival of a formalist *découpage*. Rather, it elicits a structuring of a superior order which restores the abstraction of the preceding structures, in a certain manner denying them, but also integrating them through the construction of the linguistic object on a new plane.

4.13. This is the problem of a structural semantics. I shall only say a little about it, in order to show just the dialectical movement of the formal *découpage*, and the obstacles which paradoxically condition its progress. Just as there is a scientific phenomenology of the linguistic 'sound', phonology, a scientific phenomenology of the 'words' must also be conceivable, which would be above all a morphology, whose element would be the morpheme. But the idea of morpheme is far from being as detailed as that of the phoneme. There will be attempts to define it as the smallest unit of meaning capable of appearing in different sequences. But here there is no physico-physiological criterion to define the very terrain upon which one can effect the structuring. One can, of course, try to apply roughly a statistical analysis by measuring the tendency of an element to enter new combinations in terms of frequency (cf. for example, Bolinger 1948); but this measure of semantic vitality obviously presupposes some prior indication of the consistency of elements as bearers of *experienced* meanings. Just as phonetic analysis addresses itself to a physico-physiological given, so morphological analysis rests on a psycho-social given. Two directions are available for a semantic *découpage*. One is the traditional path of etymology, which distinguishes elements in terms of the history of language, but can not furnish a description of the relations between actually functioning elements in a given state of the language. The etymological *découpage* of the English word 'disease' into 'dis' and 'ease', of the French word '*avoué*' into '*a*' and '*voué*', obviously does not correspond to a currently effective semantic structuring. The other path is that of a resolutely synchronic analysis resting on the idea of an active semantic constituent for the majority of subjects speaking a language.

Thus one distinguishes between the etymological component on the one hand, and the semantic 'formant' and the residue, deprived, at a first approximation, of significance and autonomy on the other. (The 'cran' in the English word 'cranberry', for example.) But the analysis then becomes delicate, continually risking a slide into verbalism and arbitrariness, unless it rests on the lived experience of the linguist. Bolinger has shown this (Bolinger 1948; 1950) by sketching some morphemic analyses pushed to absurdity: the English morpheme /erg/ (energy) and its variant /irk/ (meaning 'to counter energetically') can be found in w/ork, k/irk/, and cl/erk. It is clear that such considerations mark the degeneration of the structuralist point of view into a classificatory one. Here the analysis is reduced to the psychology and sociology of language, and is in danger of being lost in an ineffectual formalism. But this limitation, rather than condemning it, provides its justification. It exhibits all the more the relative and dialectic character of formal *découpage*, which constitutes the scientific object at clearly determined levels, without ever equating itself with a phenomenology of the *thing*.

DYNAMICS OF LINGUISTIC STRUCTURES

4.14. I have just emphasized what I call the dialectical character of form in the constitution of the linguistic object. Its introduction elicits in effect a stratified construction which relativizes each of its levels and mobilizes them into a system which is never finally completed. Let us now move on to a second aspect of this dialectic of forms: their capacity to provide the basis of a theory in evolution.

It cannot be denied that the Saussurian conception of a synchronic linguistics is offered above all in opposition to classical historical linguistics. But diachronic reality is by no means ignored or rejected outside of linguistics. On the contrary, it is on the basis of a synchronic description of systems that one must be able to redo the history of linguistic facts; it is no longer a fragmentary history that must be established. Classical phonetic history, for example, contented itself most often with describing the evolution of a sound in a language; historical phonology sets for itself the task of describing the successive systems of phonemes through time. As Jakobson and Halle wrote, with reference to phonetic evolution:

> The decisive factor in phonemic changes and in the diffusion of phonemic phenomena is the shift in the code ... The motor and physical aspects of these innovations cannot be treated as self-sufficient agents, but must be subjected to the strictly linguistic analysis of their role in the coding system (Jakobson and Halle 1975, p. 64).

In what direction is the structural explanation of a dynamic of languages oriented? It seems that two types of such attempts have appeared up to now. A brief examination of them will enable us to better appreciate the dynamic aspect of the formalist *découpage*.

4.15. An example of this dynamic can be seen first in an attempt to express in structural terms the ontogenesis and the pathological dissolution of language. The systems elaborated by phonologists should serve in effect as a framework for the numerous observations by psychologists concerning the acquisition of language by the child. Jakobson and Halle have sketched such an application, and we shall treat only a simplified version of it.

(1) The first sound to appear in the child's language would be the syllable /*pa*/. According to the authors, it constitutes in effect the syllabic element *par excellence*, composed of the optimal consonant and the optimal vowel. The phoneme /*p*/, is occlusive and diffuse; in it the sound energy is deployed in a very short time and distributed across a wide band of frequencies, and it is opposed diametrically to the vowel /*a*/, which is compact, and whose sound energy is concentrated on a restricted range of audible frequencies.

(2) The archetypal consonant /*p*/ is differentiated according to the tonality which opposes it, as a grave to an acute consonant /*t*/. This tonal opposition is realized naturally, the authors note, in the consonantic domain, where energy is spread over a wide range, rather than in the compact domain of the original vowel.

(3) In the vocalic domain the compact/diffuse opposition manifests itself in the appearance of a diffuse vowel.

(4) The two domains are differentiated from each other in a second way: the two diffuse consonants are opposed to a compact consonant /*k*/; and the diffuse vowel splits, according to tonality, into a grave /*u*/ and an acute /*i*/.

Finally, the successively constituted oppositions can be schematized according to the following diagram.

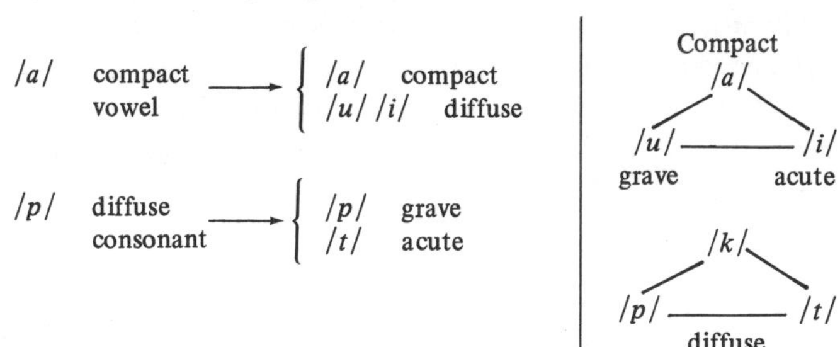

Two fundamental axes of opposition are apparent, that of compactness and that of tonality. According to the authors, they are present in all phonological systems. The authors add that if the opposition remains linear, the consonantic is effected along the axis of tonality, and the vocalic along the axis of compactness.

Their account of this dynamic of structures suggests two remarks. In the first place the authors insist on an *order* in the appearance of oppositions, which is that of the inclusion of successive schemata; this order is analyzed in greater detail elsewhere. For example, the opposition between compact vowel and diffuse vowel cannot precede the opposition between grave and acute consonant; and the differentiation of diffuse vowels in the palatal and the velar precedes the analogous differentiation of the compact vowels.

On the other hand, an idea of *optimal* realization of an opposition dominates the interpretation of the schemata. It is by attenuating, so to speak, these requirements of optimal contrast that the system of phonological oppositions is enriched. For example, the appearance of liquids, by combining the characteristic spectral trait of vowels (a well-determined formant) and the reduction of energy which characterizes the consonants, dissociates the consonant-vowel pair into two autonomous oppositions: vocalic/non-vocalic, and: consonantic/non-consonantic.

> While the consonantal feature, reduction of energy, is optimally represented by the stop which tends toward a single pulse, the non-vocalic feature, absence of sharply-defined formant structure, is optimally manifested by the strident consonant which tends towards white noise (Jakobson and Halle 1975, p. 56).[7]

By drawing the two traits of non-vocality and consonantism out of their original assimilation the appearance of liquids makes possible new combinations, apart from the optimal positions. The optimal consonant being the mellow occlusive (which manifests the most noticeable reduction of energy), the strident occlusive, or affricate, represents henceforth a possible deviation, just as the mellow consonant [*continue*] is separated from the optimal nonvowel, the strident consonant [*continue*].

The structural description of phonemes thus serves here to outline their differentiating function, their information-value. Thus, the description traces the frameworks of a dynamic by outlining the necessary conditions of the successive differentiations, which correspond to the comprehensive modifications of structures, whose common thread would be maintenance of an optimum of oppositions in situations of increasing complexity.

The same method can clarify the process of deterioration of language in aphasics: a process of 'dedifferentiation' and return to the crudest optimal oppositions. (Cf. for example, Ombredane 1951; Goldstein 1948.)

4.16. When no longer applied to an ontogenesis but to a phylogenesis of language, structuralism tends to introduce a dynamic perspective in an analogous fashion. Synchrony is then presented as a preparation and priming of the diachronic. The problem is one of bringing to light, in a linguistic system defined by its structure, the conditions of change, which is then the transition to a new structure. The leading idea here is that this evolution is not an assemblage of elementary changes involving isolated fragments of the language, but a global transformation of the system, a shift in its internal equilibrium. It is in fact the notion of an equilibrium of structures which dominates an attempt like Martinet's (in *Économie des changements phonétiques*, 1955). The coordinated ensemble of oppositions which constitutes a phonological system is considered as a sort of network of agreements, expressing the provisional equilibrium of tensions born in the conflict of the differentiation of phonemes. For each of them there is a 'field of dispersion' in the system, that is, a zone of phonetic realization that cannot encroach on those of others without occasioning a recasting of the entire network. The occlusive /*k*/ in French is realized as a velar (in front of back vowels: /*ka*/) or as palatal (in front of /*i*/) without danger of confusion; a sufficient 'zone of security' thus exists towards the front part of the mouth between /*k*/ and /*t*/ so that each of the phonemes fulfills its distinctive function. But if this zone is too narrow, and the opposition of the phonemes is prevented, a new demarcation is demanded which brings in its wake a cascade of transformations. Martinet gives a good example of structural genetic explanation with regard to the vocalism of a Savoyard patois (Hauteville). He postulates an initial structure involving four 'orders' (distinguished according to their degrees of opening), and three 'series' (distinguished by their points of articulation). The nasals complete the system according to the following schema:

TABLE I

closed\	front							back			
1	*i*			*ü*			*u*				
2		*e*		*ö*		*o*		*ẽ*			
3			*ɛ*	*	⊗				*ɛ̃*		*ɔ̃*
4				*a*						*ã*	

The incomplete order No. 3 (* corresponds to /*æ*/ and ⊗ to /ɔ/) is insufficiently differentiated, because of the weak opening opposition, from order No. 4, where the oppositions of the series are themselves neutralized; it would therefore have disappeared, the opening opposition being relayed by the front/back opposition, in a unique order: (*æ*, *a*),

TABLE II

1	*i*			*ü*			*u*
2		*e*		*ö*		*o*	
3			*æ*		*a*		

In the isolated nasal system /ɛ̃/ "exercises a pressure towards the base", cancelling out the opposition with /ɛ̃/, which is denasalized and thus reappears in the oral system. In this system, order No. 3 is consolidated, by opposing to /ɛ/, the variants of /*a*/ which pass to /ɔ/. Whence comes the final, better integrated system:

TABLE III

1	*i*			*ü*			*u*			
2		*e*		*ö*		*o*				
3			ɛ		ɔ			ɛ̃		ɔ̃
4				*a*					*a*	

Thus, the method consists in describing the successive states of equilibrium in which the asymmetries and the insufficiencies of relation appear: more than a dynamic, this is a 'comparative static', to borrow an expression from the economists.

Such an enterprise, whose interest and novelty is apparent, raises two essential problems. First of all, there is the idea, still vague, of the degree of *integration* of an element into a structure. One perceives intuitively the difference between the system of Table I in which there are empty boxes, and that of Table III, compact and without lacunae. But is it not advisable to specify and generalize this quality of completeness? If the degree of equilibrium of a structure must be a determining index for the dynamic, an exact analysis must certainly provide a definition of it. But such an analysis no doubt assumes that the structural formulation is subject to rigorous rules, that is, in the final analysis, it is molded into an axiomatic framework. Thus, just when there is a dynamization of structures, we can see this extreme requirement for formalism paradoxically heave into sight. We shall not examine this.

But the enterprise in question suggests to me a second remark. If the structural description leads to a comparative statics, which alone is capable of elucidating with exactitude and rigor the object of a linguistic science, Martinet does not assume that it must form "a set of recipes, permitting one to explain everything on the basis of anything" (Martinet 1955, p. 191). At another level, the authentically dynamic factors must take into account structural modifications. It seems that Martinet has above all brought back into play a principle of economy, referring to a physics of communications. For a given number of phonemes, there is a structure which minimizes the number of pertinent articulatory types, and the linguistic evolution would depend on the continuing antinomy between the needs for non-equivocal communication and the tendency to minimize the 'cost' of our mental and physical activity. ... Thus we emerge into a complementary conception of the linguistic phenomenon, viewed this time as the material realization of a process of communication. Only an active 'phenomeno-technique' can flesh out a theory of dynamic factors, and the usage of formal thought, rigorously conducted, requires, elicits and makes possible the birth of an applied knowledge.

'LANGUAGE ENGINEERING'

4.17. To illustrate the earliest attempts at this applied knowledge in linguistics, I shall limit myself to two significant examples. The first is borrowed from research on 'visible speech'. The point of departure here is purely pragmatic: Can we construct an apparatus to transcribe works directly into graphic signs, in such a way that an immediate visual reading of spoken language can be substituted for hearing. Apparently there were military applications in mind but at the end of the Second World War this research was redirected towards the education of the deaf (Potter *et al.* 1947). In general the work deals with the production of a *natural writing* from concrete phonetic material. The oscilloscope furnishes an image of sound vibrations, but this image is unreadable, since it cannot be structured by visual perception in the way that the sounds of speech are structured by auditory perception. The complex spectrum of sounds provides us with signals in its materiality; it is perceived as an indistinct continuity. In order to restore it to its linguistic function it is necessary to make evident the discontinuity and the generic shapes which we perceive with our ears in speech.

The idea of the inventors of 'visible speech', however independent it may be of structuralist theories, which the inventors do not mention, connects up with these theories through the unexpected expedient of a technical problem of *decoding* a 'scrambled' message. To solve it they started with the physical nature of phonemes, whose dimensions are pitch and intensity. They imagined filtering sounds, dividing the audible spectrum into a dozen bands of 300 cycles each. The partial spectra thus obtained for the same sound are then superimposed as if on lines of a musical stave, according to the pitch of the filtered components. The intensity is easily represented on the image by the value or brightness of the transcription. The apparatus of 'visible speech' thus permits the display on a television screen of the alignments of images representing an analysis of the sound spectra, where the different traits which describe the phonological oppositions must appear more or less clearly; the vowels are shaped by horizontal lines diversely distributed according to the pitch of their formants; occlusives are marked by the brusque interruption of the spectrum, accompanied or unaccompanied by a bar for vocalization, depending on whether or not they are voiced or mute.

Of course, it was not structural analyses that guided Potter, but the rather more summary classical data of experimental phonetics. In this way he lists six characteristic forms, corresponding to six types of linguistic sounds: voiced and unvoiced occlusives and fricatives, vowels, and 'compound sounds'. It is thus upon an essentially empirical analysis that the structuring of 'visible sounds' rests. Nevertheless, one can extract a morphology distinctive enough to be taught and learned in a way that will enable one to *read* speech at a reasonable speed of talking; and with no more equivocation than is occasioned in a telephone message transmitted on a bad line. Such an enterprise is eminently instructive in that it shows how the idea of a purely structural definition of the object can have direct results in a world of machines. Sensed, qualitative determination is outstripped by a successful transposition of the linguistic object from the acoustical domain into the visual domain. This transposition is already attained, one might say, in phonetic writing. But writing is based on an immediately experienced *découpage* of the sound message, and does not touch our faith in the simplicity and the autonomy of the 'sounds' of language: it remains on the level of a psychological *synopsis* of sensory messages, and belongs to a naive proto-scientific, proto-technical *praxis*. On the other hand, by audaciously introducing the machine as a means of transcription, 'visible speech' separates the linguistic object from experienced meanings. By taking the linguistic object radically out of its element, so to speak, it confers on its structural characteristic the autonomy which a true objectivation requires.

However, it will be noted that, once the first step is taken, Potter and his collaborators return to an insufficiently developed concept of signal, and they let themselves be guided in their morphological analysis by a purely physical theory of structures. In order to conclude this ingeniously set up research, they should clearly have taken account of the much more fruitful and adequate theory of phonological oppositions, renounce discerning types of *sound* through images, and looked for groups of traits which characterize *phonemes*. A reformulation of Potter's morphology could be undertaken in this perspective, which really would dissolve its remaining arbitrariness, and enable it to comprehend [what Pascal called] "la raison des effets."

The authors themselves furnish us an index of this need to extend their enterprise, when they examine the images of foreign 'sounds'. Several charts are given over at the end of their work to recording examples taken from the most diverse languages. And they note with surprise that, contrary to what one would have expected, the sound images of Chinese or American Indian languages are on the whole not different from the images of English or its related languages. Potter's morphology would be applied to them apparently without too much difficulty. Now this should not really be surprising since the principle of this morphology is essentially *phonetic*, and, from the physical point of view, the phonetic material of various languages is relatively homogeneous. A really effective differentiation of languages could be effected only from a *phonological* point of view, since these are the systems of opposition and substitution that characterize each language by its employment of phonetic material. Thus it would no doubt be very artificial to apply directly to all languages the morphology of 'visible speech' which is empirically adapted to the phonological structures of English. Only a morphology based explicitly on phonemic traits would make possible the utilization of 'visible speech' as a method of universal natural writing.

It is clear that the apparently most formalized structural analysis of the object remains the indispensable condition of progress in applied knowledge. This remarkable and, in a large measure, successful, attempt, which we have just described, stopped in mid-stream, through not having sufficiently exploited this analysis.

4.18. Still more significant and spectacular is the example of machine translation.[8] Here we are faced with an attempt at a radical analytical treatment of a complex operation. The machine which transposes a body of information out of one system of codage into another must be conceived as a function of the respective structure of both languages and their relationships. Its creation requires an analysis of language in a new, apparently entirely pragmatic, perspective but one which can profoundly transform the point of view of 'speculative' linguists. One might might almost say, without exaggeration, that the classical problem of the analysis of grammar is to the problem of machine translation what Aristotle's physics is to Galileo's, what speculative knowledge is to an effective one.

Work undertaken since 1955 has resulted not only in the construction of prototype machines (with remarkable performances), but also in attempts at linguistic analyses undertaken on new bases. In this domain the engineer continually calls upon the linguist, and indeed the logician. We find here a phenomenon which is completely characteristic of the contemporary history of knowledge: the machine *can* translate, that is, it possesses all the means necessary for the realization of the required analytical operations; but it can only do it if a suitable program guides its procedures. Now, the establishment of such programs requires a development of linguistic structures in a new direction. Neither the historical linguistics, nor even structuralism can give rise to an analysis of language immediately assimilable by the machine. We are now beginning to see the real relation between man and machine, and at the same time the profound nature of the human sciences as *applied* knowledge. The concept of a *technical unit*, which we shall develop shortly, is revealed here as an integrated and autonomous object of science. The effective use of the machine requires an adequate determination of the human phenomenon as object of science, the pragmatic requirement is blended with the very development of scientific thought.

I shall naturally leave aside the technical details and the results already obtained in the field of machine translation. I want only to underline the already discernable epistemological consequences of the enterprise, as an example of the relations between formal analysis and action.

4.19. I shall limit myself to three groups of remarks concerning the *general schema* of the operations of translation, the *process of analysis* utilized, and the conception of the *relations of the syntax and the dictionary.*

1. *The General Scheme of Translation*

In the present state of the most advanced research, machine translation proceeds in two phases; first of all, an analysis of the text to be translated, leading to certain coded results; each element is explored, identified in its semantic content and syntactic function. This operation is undertaken entirely in the 'input' language. Each element is thus reduced to an index of semantic radical, an index of syntactic class, indexes of grammatical characters (like number and gender), and indexes which signal the possibilities of homonymity [*polysémantisme*] or idiomatic expression. It is on the basis of these analyzed themes that the second phase is effected, the construction of sentences in the output language. The preparation of a translation program thus presupposes an analysis of the means of expression of two languages; the most significant feature here is that the formal analysis of one language is oriented by the characteristics of the other and that in fact a completely new *comparative* optics is developed. Thus the idea of a universal grammar is revived and transformed, but totally shorn of all metaphysics, agreeing finally with the truth of the concept. Carried away by mythological élan, one could sometimes speak of a language proper to the machine, which serves as an intermediary between the input and output languages. In reality, as the Russian Andreev (in a paper (in Russian) presented at a conference on machine translation (May 1958), cited in Delavenay (1960)) has pertinently noted, it is more precisely a question of the creation of a *metalanguage*, capable of describing linguistic facts of two languages and their relations, in a way that permits the formulation of a program of instructions to be fed into the machine. Formalization takes on here its concrete sense of instrument *which makes possible for the intelligence the construction and the mastery of a machine.*

2. *The Process of Analysis*

Of course, the microfunctioning of the translator amounts to a series of dichotomous choices, conforming to the fundamental law governing all informational processes. The instructions for a program of analysis – or of final synthesis – thus express themselves in the form of Boolean propositions directly realizable by the circuits of an electronic computer. Consider an example borrowed from the program of translating Russian into English, as studied by Oettinger (1955). The machine must decide if the letter '*ŭ*' is an ending playing no role in the semantic radical. The study of declensions and conjugations leads the author to formulate the following Boolean proposition, whose 'step by step' verification by the machine determines a positive response:

$$\grave{u}_1 \cdot [\{\sim e_2 \cdot \sim u_2 \cdot \sim o_2 \cdot \sim \text{ы}_2\} \vee \{u_2 \cdot n_3 \cdot (a_4 \vee e_4 \vee \text{я}_4)\}].$$

The symbol x_i means: "the letter x is the i-th from the end of the word."

The same is true for all the elementary operations of the machine, which are thus not differentiated, in their infrastructure, from other information processing machines being used to calculate or realize any complex program whatsoever. The originality of the translator comes exclusively from the analysis of language employed and not at all from the means to which the translator has recourse.

3. *Dictionary and Syntax*

This linguistic analysis, at present still tentative, consisted first of all in the radical reduction of a text to its semantic aspect. The earliest American attempts took the form of a word-for-word translation of roots. The preparation of a translation program then consisted in delimiting the 'radicals' which the machine can recognize and compare to the complete words of the text, thus establishing a dictionary. This first approximation to a language connects the considerations developed from a purely abstract point of view in Chapters 2 and 3. But that is only an imperfect conception, and the translating machine, in order to grapple with languages of a complicated syntax – like Russian, French, German – cannot confine itself to a continued dissociation of dictionary and syntax.

The exploitation of syntactical information is realized, for the Indo-European languages, by the analysis of endings which the machine detaches from radicals, and by the eventual examination of the remainder of each word. The comprehension of a language assumes in effect a constant interference of lexicological and syntactic levels, an interference still more and more required by its use. Thus the machine will find in its dictionary not only structural indications within the input language, but also correlative *tectonic* orders for the construction of the sentence in the output language. The problem of linguistics that this poses is thus essentially that of the *découpage* of invariant elements and structures, recognizable by the machine, and of the transposition of these invariants into the constructor schemes. The information transmitted by the input language must not trigger 'thoughts', or actions in a world of objects directly, but rather activities of symbolic construction. It is the correspondence of the two systems of expression which determines the machine's program. Thus, if the development of such a mechanism assumes and elicits a formal analysis of language, it is from a new point of view, very different from that of traditional linguistics. Such a standpoint would somehow lead to a *comparative* and *dialectic* structuralism, as linguistic forms must be defined in their relations to those of another language, and in the perspective of a transposition.

Thus the technical progress of translation machines will be correlative to theoretical progress in the conception of language and in the application of the most formal kind of thought to language.

4.20. The preceding observations were intended to show, in the very extension of formalism, the appearance of a pragmatic analysis. We shall conclude this chapter by examining in a still new and unusual field the emergence of an applied science.

'Operations research' was born of the conditions of modern warfare; timidly undertaken in the First World War, it was widely developed in the Second. The utilization of enormous means, and the manifestly economic character of the key factors of war, gave rise to the idea of a scientific treatment of military operations, which should provide commanders with the elements needed for rational decisions. It was originally in a very abstract and in large measure still speculative form that operational problems were approached. During the Great War a British officer, Lanchester, studied the advantages of the concentration of forces on a very summary mathematical model of the modern battle where a 'rate of exchange' measuring the relations of the average enemy losses appears as a unique parameter.[9] In such an attempt one could see only the episodic and otherwise rough application of habits of scientific thought to the complex phenomena where man is involved. What is more, the on-going practice of engineers has no doubt always involved attempts to formulate problems of this type. But until the Second World War, this was only a side-issue of knowledge: operations research developed and assembled a body of methods and knowledge from what had previously been only sporadic practices, stimulating the formation of specialties, learned societies, journals . . . and consumers. The development of operations research certainly poses a psycho-sociological problem whose data are closely tied to technical progress and to the economic and social conditions of our time. Leaving to the historians of science and technology the study of this aspect of the new discipline, I shall only attempt to show how, from its still indistinct object and confusion about its methods, there could arise an at least partial renewal of the science of human behavior.

In its present form, operations research embraces a set of problems that extend to all areas in which human beings in any way hold sway. It was defined in 1950, in the first anthology of papers treating it, as "a scientific method of providing executive departments with a quantitative basis for decisions regarding the operations under their control" (Morse and Kimball 1950, p. 1). Under these circumstances, the new discipline was to constitute a link between practice and most of the sciences of man and nature, to the extent that they converge in the domination of phenomena. One suspects that it is not this vague and universal equality that is the basis of its singularity and its interest. The originality of operations research derives from what it takes as the object of the *organizations* in which man plays a role, and by means of which he connects his own system of reactions to the natural systems or what he himself has set up. The natural sciences study physical, chemical, biological systems considered in their relative autonomy; the traditional human sciences take for their theme man considered as a system in itself; the object of operations research is the complex defined by G. Simondon as the 'technical individual' (Simondon 1958). Clearly, not the technical individual who is a machine in our industrial civilizations, but the technical individual of a higher order who includes the machines themselves and their conditions of concrete functioning. Technological progress is expressed in terms of the improvement of machines, becoming more and more autonomous and integrated but it has as a counterpart the increasingly explicit construction of technical macro-units: organizations in which man no longer acts as a bearer of tools nor even as a 'bearer of machines', but as an administrator and advisor of a team and its equipment – and too often, no doubt, a piece of equipment and *its* team.

Thus, the most general definition of operations research might be: *the science of the conditions of decision*, that is, a theory of human action with respect to its aims to dominate and control the course of the world, and the means that it can provide to do so. The down-to-earth and often narrowly pragmatist quality which initially characterized operations research would not let us mistake the significance of the new point of view upon which it is based. Such a point of view presupposes the rational examination of two problems: first, the *dècoupage* of the complex in question according to the strategic relations of its variables, that is, according to the possibilities of utilizing them as means of action; second, a clear and sufficiently precise definition which the organization is to attain, and the functions of the variables that one wants to maximize as much as possible. Only political economy had already authorized, under its traditional form, a perspective of this type, though reduced, at least in appearance, to a particular domain of phenomena and a region of specific values.

It would be wrong to conclude from these remarks that operations research is a normative discipline, in the sense of an organon or a body of recipes. It is not only a matter of furnishing in each concrete case a scheme for rational action; for, at the horizon of this immediate utilization there appears a determination, a new description of human behavior, which is always to some extent a process of decision or choice. Beyond the as yet somewhat rough frameworks in which operations research defines optimal actions, one can glimpse models of action in general, for human conduct can only be described scientifically in relation to criteria of optimality; the difficulty is to formulate them, without being content to live and understand them, and this is precisely what operations research is progressively discovering.

Its current field of problems is admittedly still limited and too heterogeneous to enable one to outline clearly an initial articulation of the field of action as an object of science. It does seem possible, however, to discern three main lines of research: 'programs' concerning the distribution of resources and factors among diverse places or diverse consumers; the study of 'waiting queues', which aims at determining the optimal conditions of the flow of a random flux of 'clients' waiting for a service; and the diverse competition which depends directly on the game theory of coalitions. This is a purely pragmatic classification, one which is for the most part accidental, i.e., determined by historical circumstances. It will be noted that the first two subjects stem directly from empirical problems: the second has its origin in the work of a Danish engineer on the flow of telephone communications (Erlang discussed in Brockmeyer *et al.* 1948); the first is related to the work undertaken during and after the war in order to rationalize the distribution of the factors of production and transport (Hitchcock 1941). The third originally appeared in Emile Borel's (1921) and John von Neumann's (1928) work in pure mathematics before attracting the attention of economists. Thus it is clear that in these fields, pure science and 'applied' science can exist only in a symbiotic relationship.

This is no question here of presenting a detailed analysis of any of these themes, whose shifting borders furnish no certain guide for an epistemology of this new scientific spirit. Instead I shall limit myself to presenting a summary of schematic examples, by highlighting the mode of *découpage* of the phenomenon on the one hand, and the transition from the pragmatic schema to strictly explanatory schema on the other hand.

THE THEORY OF QUEUES

4.12. Consider a flow of 'clients' arriving at a point where they are provided with a 'service'. This is a very general situation in all human organizations, and is a simple example of a technical unit: automobiles in front of a red light, airplanes waiting for the order to land, telephone calls converging on the telephone exchange, broken machines awaiting repair, etc. ... The operational analysis of the phenomenon emphasizes two concurrent processes: the distribution of arrivals, and the distribution of durations of service. Various hypotheses can be formulated about these phenomena, in themselves complex, which generally appear to be *random*. Thus, for each of them three factors may be defined which permit one to describe the state of the system: first, the distribution of the probabilities of duration of service (or of intervals of arrival): $S_{0(t)}$ defined as the probability that the service will last longer than t units of time; second, the average duration of service (or the average interval between two arrivals): T_s, which can be shown to be related to the first factor, since

$$T_s = \int_0^\infty S_{0(t)} \cdot \mathrm{d}t;$$

the distribution of the probability of the continuation of service (or of waiting time): $V_{0(t)}$ defined as the probability that *at any given moment*, the service will be extended beyond t. In the simplest case that comes to mind, $V_{0(t)}$ is identical to $S_{0(t)}$, that is, at any moment, the probability of waiting $\geqslant t$ is equal to the probability of an interval $\geqslant t$ that has passed since the last arrival. The events, arrivals, or completions of service, are produced completely at random in the course of time, and a rather simple calculation shows that their distribution is the classical Poisson, or exponential, distribution characteristic of 'rare' events.[10] A large number of phenomena in our society satisfy schema of this fashion.

On the basis of such hypotheses, the choice of critical or strategic parameters is governed by the perspective of control and domination of the phenomenon. For example, the probability of a queue of n clients can be calculated, and one can show that this probability depends in a simple way on the relation between the average interval of arrival and the average duration of service. This is the 'parameter of utilization'[11] of the technical unit which will serve as a pivot for calculations determining the magnitudes that can be used to guide the management of the system.

The choice of such an example of operations research is no doubt somewhat surprising, given the perspective in which we have placed ourselves. One might conclude that such an investigation has nothing to do with a science *of man*, since all intrinsic [human] significance is absent from the phenomenon; one perceives here only random flows, and the more or less stable states that result from their concurrence. But it is precisely at this very rudimentary level that we must study the *découpage* of the phenomenon in order to understand how meaning is introduced. The phenomenon of queues depends on the *sciences* of man, not to the extent to which it gives rise to a psychological analysis of the motives, individual or collective, which can determine the formation of a flow of 'clients', but first and foremost to the extent that it constitutes a set of controls, a technical unit, susceptible of management. It is from this global point of view that a method of rational exploration and manipulation effectively imposes itself on a domain. No doubt, the natural sciences have given rise to techniques of intervention, with which they are increasingly interdependent, since the time of Archimedes, Descartes and Galileo; but their object has not been the system of intervention itself, the technical unit, human action. In natural science the *découpage* is effected in such a way that the autonomy of a natural system becomes evident, conforming to a legitimate provisional abstraction. If one wants to admit a distinction in these terms between the two types of science, it must be concluded not that the methods of the natural sciences invade the sciences of man, but that the nascent science of human action invests the natural sciences, and that the abstraction of the natural system finds itself reduced to its true proportions in the presence of the concrete complex that constitutes the technical unit, an object still far from the new science. This is a strange and prodigious deepening of the Aristotelian dialectic of Φύσις [nature] and Ψυχή [soul].

THEORIES OF LEARNING [*APPRENTISSAGE*] AS DYNAMIC GAME

4.22. We shall now consider a very different example, in which the transition from a still very fragmented operations research to a tentative explanation is to be noted clearly. The very same steps which led to the definition of the factors of rationalization in a collective organization can be used to trace the framework for the explanation of an individual phenomenon, considered in itself as a sort of technical unit, in concrete relation to an environment. One cannot talk, of course, of operations research strictly speaking, in reference to modern theories of learning; however, one can clearly see the kinship of conception which links them to this discipline, and it is significant that a collective work like *Decision Processes* (Thrall *et al.* 1954) includes both an article like Marshak's ('Towards an Economic Theory of Organization and Information' (1954)) which relates to operations research, and articles like that of Flood on 'Game-Learning Theory and Some Decision-Making Experiments' (1954).

The phenomenon of learning is envisaged here as an adaption of the reactions of a subject to a flow of random events. This adaption is a game, in the game-theoretical sense, between the subject and 'nature'. But one of its peculiarities is that as the player does not know whether 'nature's' strategy depends on his own reactions or not, he himself modifies his own strategy to the degree to which he meets with successes and failures. In the economic theory of games, the problem is to determine the rational course of action, in the psychological theory of learning games, the problem is to describe the model of behavior whose evolution will lead to a successful strategy, one that accomplishes the learning. Thus there is here a transition from normative methodology to an explanatory and experimental methodology.

The elementary model used by Flood[12] is for a *rat*. A rat must choose between two forms of behavior, both of which are sometimes rewarded and sometimes punished. The reward and punishment are randomly distributed in such a way that each form of conduct has a particular probability of being rewarded or punished. Were he a good mathematician, the rat would get only a little help from game theory, not, as Flood affirms, because there is no game matrix in this case, but because the solution of this game is a mixed strategy which consists in giving both choices an equal chance.[13] He says:

> Nevertheless, a rat or a human found in this situation does behave in some fashion, and our scientific problem is to explain and predict actual behavior as well as possible (Flood 1954, p. 145).

The orthodox strategy following von Neumann is static; on the other hand the behavior of a subject, faced with randomness, presents itself in this case as tentative search. The hypothesis of learning is that the rat modifies his behavior on the basis of past experience. More precisely, each reaction is considered as involving a choice between the two 'pure' strategies offered, each of them being weighted with a probability coefficient that depends on the past. The tth action, for example, corresponds to a drawing by lot from an urn in which tickets prescribing behavior A are n times more numerous than tickets prescribing behavior B. But this mixed strategy is supposed to be modifiable by the subject in passing from the tth action to the $t+1^{\text{th}}$ action, as a function of the result of the tth action. In mathematical language, a mixed strategy is here a vector $P_{(t)}$ of two dimensions[14] whose components are the probabilities, $p1_{(t)}$ and $p2_{(t)}$, attributed to the two alternative courses at time t. Its modification is assumed to be conveniently described by a linear transformation, whose operator is then a squared matrix M^i. As the transformation to be attained depends at the same time on the choice made by the subject in conformity with the previous strategy, and on its success or failure, one should therefore construct two pairs of matrices, the operators for success S_1 and S_2, and the operators for failure E_1 and E_2, whose respective application depends on the results of the behavior in t. One would thus write:

$$P_{(t+1)} = M^i P_{(t)}, \quad \text{with} \quad M^i = S_i \text{ or } E_i.$$

The parameters defining the matrices are estimated on the basis of experience, assuming the model is valid, and the problem of the description of the rat's behavior then becomes one of calculating the evolution of the vector $P_{(t)}$, and in particular of determining its asymptotic behavior, which must characterize the success of learning. But the mathematician knows very little about this evolution; thus he must make use of a sort of abstract experiment, now rather common in the field of stochastic theories: a computer programmed to make random choices as postulated by the model will realize, so to speak empirically, the chains of reactions depending on the strategies and matrices, as well as the distribution of failures and successes determined by the strategy of 'nature'.[15]

It is observed that the asymptotic behavior of the 'rat' approaches the pure optimal strategy (in the Bayesian sense) which maximizes its exacted gain:[16] everything happens as if the 'rat' finally guesses the mixed strategy of 'nature'; he reproduces learning behavior. It will be permissible to compare his behavior to the real behavior of a man, and even to set up a game between the latter and a 'rat' in analogous situations. Finally, the model constructed here can serve as a reference for an analysis of behaviors, considered from the point of view of this elementary technical unit consisting of the psychic organism of man in the presence of a random flow of natural events. Such a program avoids the mechanistic character of which it might be accused by an observer too quick to draw conclusions. For the model does not at all introduce two mechanically linked systems. It postulates a structuring of the situation by the organism, and the 'discovery' of the optimal strategy determines a qualitative modification, a restructuring, henceforth stabilized, of the technical situation. Equally, the model escapes the accusation of normativity, since the theory does not directly assimilate the phenomenon to a rational conduct, but describes the evolution of a behavior which approaches this sort of conduct as a state of equilibrium.

4.23. Certainly it would be ridiculous to present attempts of this kind as definitive paradigms of science. We claim only to have found here some attempts at a new and fruitful determination of the categories of the human object, as well as the premises of a specific experimental method. It appears that the dominant category can be designated by the word '*decision*', and the experimental method, by '*operations research*'.

A theory of decisions is opposed to a simple theory of causes – or if one prefers, a theory of determinations, such as the natural sciences offer – in that on the one hand it introduces both a random complex, and an optimum, and on the other hand it articulates an apparatus of *information* and an apparatus of *action*. The objectivization of these two notions, elsewhere charged with experienced meaning, characterizes the present phase of science. The philosopher who wants to understand the implications of these attempts must avoid the temptation of vulgar ontologism. The new science does not reconstruct man with machines, but conceptualizes human situations as technical complexes in which both man and nature are both engaged. This is a conceptualization which clearly surpasses the simple *découpage* of phenomena, however much this latter is essential; we will see in a later chapter how this conceptualization takes on its full sense through a dialectic of axiomatization.

But the scientific revolution in the domain of mankind consists first of all in freeing itself from the naive modes of *découpage* transmitted by ordinary language. Human events, taken at the level of experienced meanings, can give rise only to a pseudo-science, a more or less skillful discourse which only reflects an empirical practice, even if it is raised to the level of an art. The transmutation of the phenomenon into an object is achieved by the convergence of two movements which cause forms to penetrate into the world of events. We have described the first under the name of formalist *découpage*, for it aims directly at the construction of abstract systems which it studies apparently for themselves. The second is revealed in enterprises like operations research, in which formalization is subordinated to a perspective of action. But we have shown that the two movements presuppose each other and rejoin, offering a glimpse of the possibility of an original discipline, which should become the science of the future.

It would be appropriate to extend this interpretation by the analysis of the central theme of this conceptualization at work in the science of man. I shall examine the treatment of *quality* in the next chapter.

CHAPTER V

QUALITY AND QUANTITY

5.1. Behind most of the criticism advanced against those who support a rigorous science of man, one finds the objection concerning *quality*. There is always the fear that a scientific understanding will overlook what seems to be most significant in the human being and his works, what is most unique and most irreducible to schematizations of any sort. Obviously one must admit that, in the four centuries since Galileo, Lavoisier and Claude Bernard showed us the way to the conquest of non-human objectivity, the physicist, the chemist, even the biologist have been able to detach themselves from *warmth* and *wetness*, from the *sweet* and the *bitter*. But psychological or social reality is still supposed to be grasped by the scientist as immediate experience gives it to us, that is, as a tissue of qualities. If in the domain of natural entities it seems easy today to think of quality as appearance – or more exactly, to admit another phenomenology, according to which the object is determined by abstract schemata which enable us to grasp it effectively, in the domain of man such an approach apparently meets with much difficulty. The view is eagerly embraced that the very *essence* of the phenomenon here is qualitative. Bergson founded his metaphysics and his dualistic theory of knowledge on this lemma. But many of the very people who are trying to build science openly share this perspective. Some sociologists have challenged the use of mathematics because it is founded on the indifferent relations of a whole and its parts, while the human totality, the total social fact is an organically and qualitatively differentiated ensemble, whose dynamism cannot be conveniently described except by the mediation of images, and in particular by that of "perpetually boiling igneous material" (G. Gurvitch 1955, p. 40).

Is this theme of the fundamentally qualitative nature of the human fact merely a pre-scientific prejudice? What is its exact meaning, and how can the various sciences of man practiced today orient themselves towards a rational treatment of the qualitative? This is the problem I want to pose now; it is the problem which brings the philosopher to the very heart of an epistemology for the human sciences.

QUALITY OF THE OBJECT AND QUALITY OF THE LIVED EXPERIENCE [*VÉCU*]

5.2. The philosophy of science, fascinated by the problems of structural coherence posed by the modern developments of physics, has almost completely neglected the category of *quality*.[1] Contemporary interest in the sciences of man requires a reconsideration of this old problem, at least an exact analysis of the presuppositions which use of the concept of quality introduces into the sciences.

The perception of quality corresponds, genetically as well as phenomenologically, to the immediate moment of knowing. But this immediacy is equivocal. Husserl, on a fine page of the *Ideas* (Husserl 1952, p. 139), nicely insisted on this duality of the qualitative: on the one hand the qualitative involves the immediacy of the thing, whose transcendence is only given in outline, always incomplete, but always grasped in a presence; on the other hand, the immediacy of what is experienced, is given as a flux, and consequently given in an incomplete way totally distinct from the former aspect of the qualitative. "Only in the form of retention or in the form of retrospective remembrance have we any consciousness of what has immediately flowed past us . . . But *this* incompleteness or 'imperfection' which belongs to the essence of our perception of experience is fundamentally other than that which is of the essence of 'transcendent' perception, perception through a presentation that varies perspectively through such a thing as appearance" (Husserl 1952, p. 140). In both cases, however, we are concerned with a perception, that is, with the *immediate* sight [*visée*] of an object, transcendent in relation to consciousness, and it is this mode of appearance of the object that we call quality. However, an over-hasty confusion between the quality of the external object and the quality of the psychic object, between the outline-quality and the tonal quality, so to speak, leads to inextricable difficulties. The outline-quality, for instance, the redness of a book cover, the stridency of a cry, deserves the name 'appearance' insofar as it is only the contingent and relative predicate of an object, perceivable in other ways. The tonality (and not just the affective tonality) of my experience of myself, while I write this, is not an appearance, but the very essence of my consciousness which passes, of the psychic object that I perceive in myelf. The Bergsonian critique (in the *Essai* [Bergson 1889]) plays on a tacit assimilation of these two types of qualities which it tends to reduce to the second type, while the associationism of a Taine tends to identify our qualities of consciousness with the first type. This assimilation is no doubt one of the secret resources of esthetic symbolization. Art in fact plays on this ambiguity by using the representation of objective qualities as if they were aspects of the experienced. By means of a truly very spontaneous transmutation, the artist evokes plastic and colored qualities, which are aimed at as such and as aspects of consciousness, thus achieving a magical appropriation of the world of things as opposed to the cognitive and technical appropriation that constitutes the work of the scientific spirit. This is why art is on this side of the distinction between idealism and realism, between representation and abstraction. In its very essence it accepts the equivocal and refuses dissociation. It is from this that the seductive force of a Bergsonian philosophy derives, which is, in many ways, a cunning thematization of esthetic activity.

5.3. But if science, as its history has clearly shown us, has the task of dissipating this equivocation and of setting up a mode of resolutely *objective* thought, it must certainly avoid this ambiguous attitude in the face of the psychic fact. It obviously follows from this that a science of human facts must proceed from a conception of quality completely different from that of the physical sciences, but in no way must it limit itself to rejecting as outside of its scope some particular object which is qualitatively given to it on another mode. Today we can begin to glimpse the relations between sensible quality and scientific schematizations of the physical object; we can begin to clarify in a specific manner, the relation of the quality of a psychic object and its structuring as an object of science: this is without doubt the fundamental task of an epistemology of the human sciences. It would, however, be fruitless to demand that the philosopher invent this elucidation all of a piece; it is the very progress of science that can suggest it to him. Undoubtedly, we are not even advanced far enough in this field to hope to provide this elucidation in the immediate future; at the least the time has come to sharpen the problem and to analyze the beginnings of the spontaneous solution that the psychologists and sociologists are bringing to it.

Having declared the danger of an overly simplistic reduction of the qualitative, let us examine the sense of quality in the realm of human facts. The most obvious feature of quality in its two modes is that it is an aspect of the *in itself*, more precisely of the *being there* (determinate being). Hegel introduces it right at the beginning of the 'greater' *Logic*, in his theory of being; quality corresponds to being insofar as it is determined, it corresponds to *Dasein* whose determination is the determination of being (*seiende*), that is, quality (Hegel 1969, p. 109). For the thought of the object, the aspect of quality is in fact the aspect of the immediateness; but it is not a frozen immediacy, a stopping and resting place for the understanding. On the contrary, scientific understanding can even be defined as the movement of thought which is neither satisfied nor halted at the point of the immediate.

> Determinateness thus isolated by itself in the form of *being* is *quality* – which is wholly simple and immediate ... Because of this simple character of quality as such, there is nothing further to be said about it (Hegel 1969, p. 111).

But

> determinate being ... in which nothing no less than being is contained, is itself the criterion for the one-sidedness of quality as a determinateness which is only *immediate* or only in the form of *being* (*ibid.*).

The thought of the object goes beyond this immediacy, this unilaterality in developing its negative aspect. Quality is then essentially grasped as limitation, or, more precisely, as *difference.*

DIFFERENCE AND SIMILARITY

5.4. It is undoubtedly through this very aspect of difference that the qualitative can be conceptualized, and gives a handle to a dialectic. Quality is the Aristotelian *διαφορὰ τῆς οὐσίας* [differentia of substances] of book Δ, but the difference only makes sense in a system of oppositions and correlations, which makes us pass from the immediate and apparently isolated *being there* to a *structure*. Phonology has already furnished us an example of this dialectization of the phonetic qualitative. Thus, from now on it is apparent that this conceptualization of quality is not necessarily effected by a transition to the quantitative, as the Hegelian *Logic* would have it. The fundamental philosophical progress of mathematics in the twentieth century is the recognition and acceptance of this nonquantitative dialectic of quality. It will be useful to dwell on this discovery since, for many of those who work in the human sciences, mathematization is equivalent to the introduction of quantity, indeed of number.

In fact, now that the successive extensions of classical algebra have led mathematicians to conceive, first, the general idea of algebraic structure, and then that of any structure at all, we can better understand the hitherto implicit movement which established the transition from qualitative thought to mathematized thought. The first step of this dialectic leads to the notion of *set*, that is a universe of objects whose qualitative presentation is reduced to just its essence of *difference*. Members of sets are abstract objects of which one wants to know only whether they are distinguishable from one another from a certain point of view, although in other respects they may be mutually substitutable as elements of the set. This is an undifferentiated difference which permits the elaboration of a mathematics. Once this reduction is completed, the initial steps of the mathematical treatment consist, on the one hand, in *joining together* these objects in all possible ways into subsets which are equally distinguishable, and whose common elements can be discerned; on the other hand, it consists in *bringing into a correspondence* the elements of two sets, which are associated in pairs until one of the two sets has been exhausted. Thanks to this it will be possible to characterize a set of abstract objects by describing, or constructing, the relations which can be established between the whole and its parts. Thus, the notion of operation appears quite naturally as the correspondence between an element within the set of results, and an ordered pair whose terms belong to the set of elements on which the operation is carried out. The distinctive properties of an operation will be described uniquely by means of the original steps already indicated. The qualitative is thus installed within the conceptual form of a *structural property* whose sense depends, not on the isolated determination of an individual object, but on a system of potential manipulations effected on a set of objects. This is the way one defines the entity 'integer': as the specific structure algebraists described with the name *infinite ring*, in which two operations intervene. Clearly, this determination of integers is not exhaustive; other properties remain free [2] within the limits of this structure, so that, at a more elaborate level of determination, other types of entities will be distinguished, belonging, however, to the same family, in that their set is structured in the same fashion. Modern geometry speaks of rings of polynomials and rings of matrices. But each new step taken to make more precise the nature of these entities is taken in terms of the structure that it expresses, a finer and richer structure which it constructs. Quality as an immediate aspect is thus mediated by passing from an intrinsically given determination to an extrinsic determination of the modalities of the difference.

5.5. We can see that this notion of difference naturally gives rise to the opposite notion of similarity and this pair of concepts is constitutive of the *concept* of quality. Similarity appears at two levels in structural thought. First of all at the level of individuals posited as undifferentiated *from a certain point of view*. The relation of *equivalence* defines in a set a subclass of undifferentiated objects, corresponding to the intuitive notion of 'qualitative identity'. This relation is of course capable of being determined by purely formal properties: reflexivity (for all a, $a = a$), symmetry (whatever a and b are, if $a = b$, then $b = a$), transitivity (whatever a, b and c are, if $a = b$ and $b = c$, then $a = c$). In this way arithmetic introduces the notion of congruence: two whole numbers are said to be congruent *modulo n* if they differ from one another by a multiple of n, and in this perspective they are considered as equivalent.[3] The abstraction of a qualitative difference distinguished from all other differences is thus effected.

But on the level of the structures themselves the notion of similarity is again introduced. Two structures are similar when a one-to-one correspondence can be established between the elements of the sets which they govern, and between their respective operations, in such a way that the result of one operation of the first sort, carried out on two elements of the first class corresponds to the result of the correlative operation of the second sort, carried out on the images of these elements in the second class. A precise definition[4] of structural resemblance can thus be given, conceptualizing the intuition of qualitative similarity.

5.6. The implications of this algebraic analysis for the human sciences are considerable. One very simple psychological example is enough: the explanation of the mechanism of the perception of symbols. Graphic or phonetic, the symbols of a language are given to us as qualities – tones or forms. Now, the comprehension of these symbols assumes a reduction of these immediate qualities such that the variations in scale, and certain variations of their quality itself, do not alter their semantic value. To understand the act of perception of these symbols thus involves conceptualizing the idea of similarity underlying their interpretation. Can one say that two manuscript representations of the letter *a* carry the same sense because they have the same graphic structure? The intelligibility of the sign persists, however, despite considerable alterations of its form. The notion of isomorphism is insufficient to give an account of this type of equivalence. In order to obtain an adequate conceptualization of the qualitative in the domain of the perception of symbols, we must add to the notion of isomorphism another idea elaborated in modern mathematics by the geometers, that of 'proximity' between objects in general, and particularly between two structures.[5] It is the intuitive and vague notion of the 'pretty close to' [*à peu près*] which must in turn be conceptualized. Modern mathematics already possesses a complex theory of the 'pretty close', of which very elaborated themes of '*limit*' and '*convergence*' constitute the fundamental instruments. A topology of abstract structures, originally constructed to describe the properties of sets of points of geometry, thus furnishes the model and the point of departure for a *theory of the 'pretty close to'*, of approximation indispensable in a science of quality. Now it is up to the sciences of man to suggest new problems to the mathematician.

It seems then that a scientific elaboration of qualitative notions consists in the transition from the *astructured* to the *structural*, rather than in a quantification. At all events, the transition to the quantitative is only one possible result of this dialectic, and it is from this perspective of set theory that one must try in any case to understand the treatment of quality as established in the human sciences. With this reservation three typical processes of this dialectic of quality are distinguishable:

(1) The transformation of qualitative data into a quantified system.

(2) The maintenance of the non-quantitative character of data in a mathematical schema.

(3) The qualitative mutation of systems whose dimensions grow beyond a certain threshold.

In our third type an application of the Hegelian and Marxist 'law' of the transition from quantity to quality is obviously recognizable. We shall see that this transition should be interpreted epistemologically in terms of a structural differentiation of phenomenological differences in the scientific object. The first two processes furnish us with excellent evidence of the still tentative character of the work dealing with formal thought as it deals with the human object; it is still a fragmentary and tenuous attempt, but already sufficiently formulated to exorcise the spectre of a brutal quantification of the unquantifiable. Here again, it is the transition from the astructured to the structural which can clarify the latent sense of these scattered enterprises.

QUALITATIVE RESPONSES AND INFORMATION

5.7. From this point of view we shall examine certain techniques recently introduced into psychology and social psychology, which are good for showing the quantity-quality dialectic in its detail and its uncertainties. In particular, one can confirm that the distinction between the first two processes mentioned above is only apparent and that, in a sense the qualitative character of the data is maintained through an initial goal of quantification.

Essentially, we shall examine the treatment of responses furnished by the members of a group to a questionnaire. The latter is supposed to disclose the presence of a certain trait or a certain attitude in the group considered; the questionnaire was set up in this sense by intuition and guesswork. But the results of the tabulations naturally furnish only a multiplicity of types of responses, a somehow qualitative multiplicity, from which one wants to cull either a score for the individual relative to the trait considered or an ordinal classification.

Take, for example, the battery of questions constructed by Stouffer to study the 'morale' of the US Army during the Second World War.[6]

(1) In general, how would you say you feel most of the time, in good spirits, or in low spirits?

(2) If it were up to you to choose, do you think you could do more for your country as a soldier or as a worker in a war job?

(3) On the whole, do you think the Army is giving you a chance to show what you can do?

(4) In general, how well do you think the Army is run?

(Stouffer *et al.* 1949–50, 1, pp. 86–87).

As the answers to these questions were dichotomized, for each individual there corresponds an 'opinion'[7] constituted by a series of yes or no answers. The problem posed by the psychologist is to produce a homogeneous significance for these different opinions.

5.8. First of all, it is worth noting the dichotomous character of the questions posed. It is in fact on the basis of this characteristic that a first structuring of the data collected is outlined. If we compare a questionnaire of this type with an ordinary interview, we can gauge the importance of this structuring. In the second case [interview] the responses are strictly qualitative, and constitute multiplicities that are completely incomparable. In the first case [questionnaire] the dichotomous opinions already involve a principle of classification; in any event their very nature involves an *a priori* determination of all possible opinions, and an elementary combinatory theorem assures us that for a battery of k questions, there are first and last 2^k types of responses, or opinions. But this sort of determination of the range of possible responses is precisely the fundamental condition for the definition of the value of a unit of *information*. One is reminded of the process of dichotomy Plato described in the *Sophist*, where we move from a question of complexity and nuance to a battery of simple questions that call for yes or no answers. But the case of the definition of the *Sophist* is special, for the questionnaire is organized into orders and is already in the form of a 'tree' [or branching system] bearing a series of bifurcations. The subjects approached by the psychosociologist in general exclude all *a priori* organization of this kind.[8] The set of questions only constitutes a given whole in which no hierarchy is outlined. The problem is precisely that of constructing an order for this as yet qualitative multiplicity, whose *finite* and *discrete* character already gives us partial satisfaction.

In contrast to this multiplicity of nuanced responses, which vaguely evoke the idea of the continuum and the infinite,[9] we find the plurality of distinct elements which constitute the types of responses, finite in number. Now, these two characteristics of finiteness and discreteness (the first of which obviously governs the second) have already appeared to us as *the most radical conditions of symbolic expression*. A language really exists when there is a choice of distinct signs in a finite dictionary. The dichotomous reduction of questionnaires thus appears to make possible the transmutation, in a manageable universe of language, of the qualitative multiplicity of empirical data.

This first reduction, however, provides the psychologist with only one condition for further development. It is not enough to be able to enumerate all the possible opinions, and to assign one of them to each individual questioned; the mutual relations of these opinions remain, at a higher level of abstraction, qualitative relations.

PROBABILITY OF RESPONSE, AND DIVISION INTO LATENT CLASSES

5.9. Lazarsfeld's and Stouffer's idea was to construct, on the basis of the differentiated set of opinions actually collected, a classification or structuring of the whole population relative to the particular trait or attitude considered. Imagine then a structural schema which is given empirical expression in the distribution of opinions collected.

It is assumed that the population is ordered into *classes* relative to the attitude envisaged. The relative significance or frequency of these classes is unknown. A class of respondents, X, is defined by the probabilities of positive responses to the different questions on the part of its individual members. In class X, a individual will respond 'yes' with the probability P_j^x to question j. This probability is of course unknown, and the theory is supposed to provide the means of estimating it. Finally the analysis which the psychologist chooses should permit us to fill the following table with probabilities which define the 'latent structure' of the group, relative to an attitude studied:

CLASSES		1	2	3	4	...	x
questions	I	p_1^1	p_1^2	p_1^3	p_1^4		p_1^x
	II	.					
	III	.					
	IV	p_4^1	...	...	...	...	p_4^x
frequency of classes		ν_1	ν_2	...			ν_x

The data obviously reduce to the frequencies of different opinions, in the group, frequencies which can be assimilated[10] to an estimate of the probabilities that any opinion actually offered in the group will be of the form $(i\,j\,k\ldots;\,l\,m\,n\ldots)$, the positive responses being given first, the negative ones after the semi-colon. Thus for n questions, there are 2^n experimental values, $p_{ij;\,lm}$, representing the frequencies realized by different opinions. How can we move from these data to the latent structure? A new hypothesis will permit us to describe the equations of the structure. Suppose that for *individuals belonging to the same class*, the probabilities of positive response to different questions are independent. This is a property that probability theorists express classically by writing that the probability of a positive response to both of any two questions i and j, is equal to the product of the respective probabilities of the separated positive reponses to i and j:

$$p_{ij}^x = p_i^x \cdot p_j^x$$ [11]

Under these conditions the probability of any opinion at all *in a class* is calculable, on the basis of the elementary probabilities p_i^x:

$$p_{ijk;\,lmn} = p_i^x \cdot p_j^x \cdot p_k^x \cdot q_l^x \cdot q_m^x \cdots,$$

where $q = 1 - p$, the probability of a negative response.

And consequently, by summation, the probability of any opinion whatever in the set of the group can be calculated:

$$p_{i\cdot j\cdot k\ldots;\,l\cdot m\ldots} = \sum_x \nu_x \cdot p_{i\cdot j\cdot k\ldots;\,l\cdot m\ldots}^x = \sum_x \nu_x \cdot p_i^x \cdot p_j^x \ldots.$$

Thus by equating these expressions to the empirical values of frequencies, 2^n equations can be obtained for the calculation of the p^x and the ν_x. We shall leave aside the mathematical difficulties of the calculation. (Cf. B. F. Green 1951.)

5.10. What exactly is the nature and the scope of the elaboration realized in this way? In the first place, it will be noted that the 'qualitative' plurality of the opinions has in no way been reduced in a brutal fashion to the limits of a numerical scale. The analysis of latent structures does not even result in *ordering* the set of opinions. It only provides a structure for the ordering for the group itself, in which the classes are distinguished according to their more or less neat adhesion to the attitude being examined. (In the example cited here, Stouffer establishes a three-class structure – good morale, mediocre morale and bad morale – defined with precision by the parameters of the table.) Each of the classes is determined by the cluster of probabilities attributed to each opinion, in such a way that the subtle and fluctuating character of actual responses is found to express in a certain sense the *aleatory properties* of the structure. Quality is conceptualized by means of probabilistic parameters. It is good to emphasize this fact, which seems to me to be a general trait of the epistemology of the human sciences. Correcting the rigid features in the informational reduction of the responses, the random parameters reintroduce under a form accessible to calculation, something of this qualitative multiplicity. The scientific schematization is obviously not in any case the equivalent of the intuitive data, no more than the equations of the physicist recreate the sensible world. But the process of transmutation stems here from the same spirit, from the same concern to construct a manageable structure, which is both fruitful and provisional.

Let us now turn our attention to the meaning of this structure, in respect to the understanding of individuals. In fact once the structure is formulated and tested[12] it in no way enables us to determine which individuals belong to which class. But knowing the opinion of an individual, one can, at the most, conjecture, within a certain range in error, that he belongs to the class which is richest in opinions of this type (the modal class). Alternatively, if one assigns to each class a relatively arbitrary numerical value (for example, 1 to the class most favorable to the attitude studied, 0 to the intermediate class, and −1 to the last class), it is possible to associate a *grade* with each opinion, and consequently to order the set of respondents in relation to the trait being studied. We need only give as *weights* to the value of each class the frequency of this opinion in the class. An opinion represented two hundred times in class 1, fifty times in class 2 and forty times in class 3 will then yield the result:

$$(1 \times 200) + (0 \times 50) + (-1 \times 40) = 160.$$

This result is the midpoint on the scale of classes, of the position of the subjects who expressed this opinion. The set of opinions is thus provided with a metricized, or at the very least, an ordered structure. But the arbitrariness of such a quantification, whose value can only be limited to particular cases is evident. The real interest of the analysis is not in this too hasty transition from quality to quantity, but rather in the structuring presented by the table on p. 94. This structure was obtained by means of two reductions, whose utility for the consitution of economic concepts I have examined elsewhere (cf. my (1955), part II, Chapter 1, Sections 15–16). The first transforms the multiplicity of possible responses into a finite articulated *informational* schema. This is a set-theoretic reduction, defining the elements in a whole: indeed eventually defining relations of 'nearness' between the elements (a step is not presented here). The second, which somehow mobilizes this rigid structure – and prepares its dynamic treatment – introduces parameters of randomness; it is a 'probabilistic' reduction.

SCALING STRUCTURE

5.11. On the basis of the preceding example, it is now possible to show how this analysis of 'latent structures' is involved in a comprehensive process of dialectization of quality. This examination will show us on the one hand the complex tendencies at work in this treatment of the notion, as well as the temptations of intemperate formalization which it involves; on the other hand, in spite of these obstacles, we shall see the general outline which emerges from these different techniques and their meanings.

The technique of Guttman scales, like the analysis of latent structures, begins with the qualitative multiplicity of types of responses to a questionnaire, and aims towards an initial ordinal structure of the population of individuals. But here one requires more from the questionnaire, for one wants each response to be *determined* by the rank, in the scale to be formulated, of the individual being questioned. To put it another way, one should be able to affirm a sort of embeddedness of dichotomous questions, such that an affirmative response to question number one is compatible only with an affirmative response to question number two; in this case only certain 'opinions' can appear, the others being excluded by this rule of coherence. The prototype of such a questionnaire would be furnished by an enquiry bearing on unambiguous data, for which there is a natural scale; but as soon as the uncertainty of the responses disappears, the interest of the questionnaire obviously disappears. Consider these questions:

(1) Are you taller than six feet in height?

(2) Are you taller than five feet, ten inches in height?

An affirmative response to (1) naturally involves an affirmative response to (2). The only coherent opinions are in fact: *yes–yes, no–yes, no–no*. The scale is here one of heights, and the rank of an individual on the scale determines his response without ambiguity.

It is this particularly favorable situation that one would like to produce or whose approximate realization one would like to test, by a questionnaire bearing on psycho-social attitudes, for which there is no process of decision analogous to the measures of height. The schematic example of Guttman is as follows. It concerns the construction of a scale of attitudes in regard to war. The questions posed are thus:

(1) Is war good?
(2) Is war bad?

40% of the subjects responded *yes* to (1), and 80% *yes* to (2). The hypothesis of coherence then permits the determination of the theoretical proportion of the different possible opinions. The 20% who responded *no* to (2) must be included in the 40% who responded *yes* to (1). The opinion *yes–no* thus represents 20% of the set. There remains 60% for whom the opinion was *no–yes* (the least favorable towards war), 20% for whom the opinion was *yes–yes* (qualified) and there were no responses of the form *no–no*, which be excluded as indeterminate, if the scale is coherent.

		Is war good?		
		yes	no	
Is war bad?	yes	20	60	80
	no	20	0	20
		40	60	100

This distribution of opinion gives rise to a linear representation, analogous to the classical percentiles of the statistician:

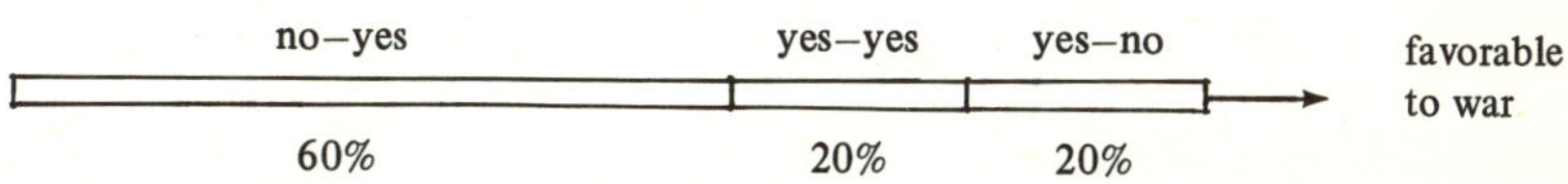

The 100 subjects are supposed to be classified with respect to their attitude, the first sixty (unfavorable) are of the opinion *no–yes*, the following 20 of the opinion *yes–yes*, the last 20 of the opinion *yes–no*. The problem for the social psychologist consists thus in constructing his questionnaire, in such a way that he can order his questions like the figure below, so that the opinions will be arrayed in a parallelogram. The technique of Guttman scales is only an elaboration of the processes and recipes which permit an approach to this organization, or a decision about the proportion of aberrant opinions which would reasonably involve the rejection of the hypothesis of coherence.

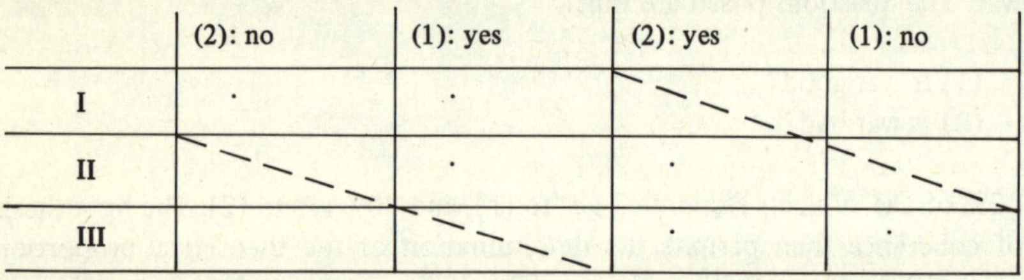

SEARCH FOR A METRIC

5.12. Here the steps towards quantification of quality reduce to an ordering, one bound, however, to a *découpage* into classes of determined frequencies, like the theory of latent structures. But Guttman pursues the exploitation of his scheme further. For the simple ordinality he wants to substitute a metrization, which will assign to each subject not a *rank* but a grade, which makes possible a comparison of the 'distances' between different subjects on the scale, and permits the assignment of a *zero* point, somehow corresponding to an objectively neutral attitude. The scale leaves us completely free to choose the grades arbitrarily provided that the order of the subjects is preserved. Guttman then makes explicit his own criterion of choice. A grade is attached to each individual depending on his 'opinion'; within each group of subjects having responded in the same fashion to one of the questions, the distribution of grades will be more or less scattered, for one question is not enough to determine the grade of each of the members of the sample. Would it not be reasonable to distribute the grades in such a way that this dispersion is minimized? Quite to the contrary, between the average grades of each of the groups corresponding to different questions, the intervals should be as large as possible, in order to obtain a better differentiation for the grades. In statistical terms, one satisfies this two-fold requirement by maximizing the intergroup variance and minimizing the intragroup variance.[13] Given these conditions the mathematician is in a position to calculate the system of grades which satisfy these requirements; however, his equations generally furnish him with as many systems of solutions as there are elementary questions, systems which are not equally satisfying but which are ordered according to the extent of the variances which they permit. The best among them constitutes the metric best adapted to the requirements, and is the only one which actually preserves the order of the scale. If one draws a graph with the ordered percentages of the different opinions on the abscissas, and the grades assigned to the different opinions by the new metric on the ordinates, one obtains a monotonic function. Similarly, Guttman draws the graph corresponding to other solutions (cf. Figure on p. 100); the curves obtained contain points where their variation changes direction. They cannot generate a metric compatible with the order of the data. But can a psychological interpretation for them be found?

THE INTERPRETATION OF 'PRINCIPAL COMPONENTS'. RETURN TO STRUCTURAL ORGANIZATION

5.13. This is the question asked by Guttman to which he gives a positive answer.

Just as mathematical constructions have often given rise to abstract entities that the physicist was subsequently able to coordinate indirectly with phenomena, so the elaboration of a quantitative structure here gives rise to abstract systems. Is it necessary to give a meaning and a role to these systems in the structuring of the experiment? Guttman and his collaborators certainly thought so, and formed the hypothesis that the second solution to the problem of a metric corresponds to a second-order quantification which would be that of the 'intensity' of an attitude. The curve representing this second solution in fact presents a unique minimum, which designates a point of 'indifference' with regard to the problem posed; on one side and the other there are subjects whose positive or negative attitudes are increasingly categorical in their positions.

According to Guttman, specific tests, constructed to estimate this 'intensity' empirically, corroborate the interpretation offered.

So audacious a reintegration of abstract elements into the phenomenon is singularly significant. It is a good idea to emphasize first of all that the integration is, in principle, *a natural consequence of the notion of structuring.*

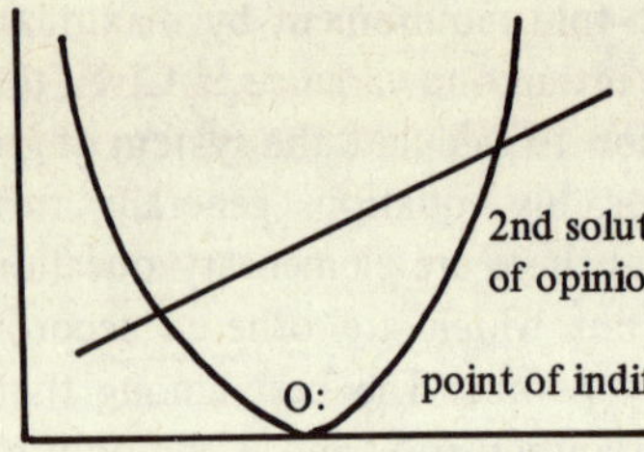

If the abstract structure that the scientist constructs is to have some sort of significant bearing, then at least certain of its mathematical properties must be susceptible of interpretation. Failing this, the whole process of reduction of phenomena to structures is only a superficial artifice. It is none the less true that the unexpected identification of an algebraic solution with a psychological trait shocks us, when that trait has not been explicitly introduced in the model in the first place. In the present case, the model is constructed on the hypothesis of the coherence of opinions, proceeding from the assumed possibility of a linearly ordered scale of attitudes. The calculation introduces a second organization, construed as an expression of the interplay of an 'intensity' of attitudes, that is, in short, a second degree attitude bearing on the original attitude. All the more reason for it to be the same for the interpretation of the other successive solutions which Guttman suggests correspond to what he calls 'closure' (the degree of conscious determination in regard to an attitude), and 'engagement' (*involution*: the interest actually accorded to the attitude under consideration). Guttman, it is true, seems conscious of this reluctance [to pursue these interpretations] and he himself comments on the relation between mathematics and psychology in order to eliminate the justifiable charge that he has drawn considerable psychological consequences from a very poor empirical scheme, simply by means of calculation. First of all, it is not in calculating systems of solutions on the basis of original data that new psychological elements like the indifference point are actually determined: mathematics itself shows the vanity of this undertaking, for it is necessary to insure the validity of the questionnaire as an adequate sample of the set of questions that reveal the attitude. It requires a direct empirical estimate of the higher order components – correlatives of different solutions for the metrization undertaken – being made by the psychologist and reported on the scale of ranks. Mathematics thus only guarantees the invariable significance of the elements thus established. "Mathematics alone couldn't determine the zero point empirically; psychology alone couldn't prove the zero point invariant with respect to the sampling of items. This successful combination of mathematics and psychology suggested it might be worth exploring the third mathematical component by psychological means . . . " (Guttman 1954, pp. 250–251).

5.14. The psychological problem thus turns into an elaboration of auxiliary tests which permit the direct estimation of the higher order components, whose intuitive meaning is conjectured. The results of these tests, reported in a scale of ranks, rather faithfully reproduce the character of the mathematically calculated components, and this is the only – unconvincing – guarantee of their validity. Thus it cannot be concluded that the undertaking has been a total success. But beyond its success, which depends a great deal on the elaboration of the data, what interests us is the very design of the process which the author has been led to develop. Starting with a desire to provide a reasonable quantification of a qualitative multiplicity, he is led to emphasize not so much the quantification itself, as the structure of the reference frame that he has outlined: he is aiming at a linear scale of attitudes, but he constructs a multi-dimensional framework, in which each subject is characterized not only by the 'content' of his attitude (his place on the original scale), but also by the different components which, so to speak, qualify it.

A profound consequence of this one-dimensional property is that it implies at the same time infinitely many dimensions – namely, the principal components. The same attitude has these infinitely many dimensions, the functional dependence on them now being expressed by linear equations . . . A scalable attitude is one-dimensional in the sense of rank order, but is a function of infinitely many psychological components (Guttman 1954, pp. 248–249).

Moreover, this infinity is practically reduced to the first terms, for the contribution of the succeeding terms becomes too weak to retain any experimental sense. The essential importance assumed by the system of components considered as a system of reference appears clearly in Guttman's remarks about the choice of an empirical function of 'intensity':

One cannot prove that one 'intensity' function is 'really' the second component and that another is not, without reference to a complete set of higher order components. There is no other way of choosing objectively between empirical psychological functions which have the same general shape (Guttman 1954, p. 247).

It is thus in this direction that the elaboration of the quantity-quality problem is to be oriented. Like the theory of latent structures, the theory of Guttman scales (which can moreover be presented as a special case of Lazarsfeld's scheme, in which the probabilities, $p_{ij}^{x} \ldots, k \ldots$ would be degenerate, taking only the values 0 and 1) moves us from the qualitative to the structured. Certainly numerical magnitudes intervene, but they are hardly more than indexes. What plays an essential role, what creates the object of research, is the structured *referential* [scheme] in which the indexes are given meaning. Now, the two attempts examined here join an older and more general one on this point, and this convergence is particularly instructive: Spearman's and Thurstone's factorial analysis in fact uses the same mathematical tools, as Guttman and Lazarsfeld themselves certainly note.

THE GENERAL THEME OF LINEAR STRUCTURES

5.15. Given a plurality of tests applied to a population, which are supposed to uncover an attitude, factorial analysis consists in constructing a rather simple system of reference by means of which the multiplicity of scores furnished by the tests will be related to the interplay of some underlying, and if possible, mutually independent 'factors'. Here again the issue is the transition from quality to structure, for the multiplicity of scores furnished for a given individual by the tests produces the largely illusory character of an *isolated* quantification of the results of a test. Should the occasion arise, the referential system of factors can be broken down into 'common' factors and 'individual' factors, but the underlying nature of the problem remains unchanged. Mathematically, the question is developed according to the process of the two-part reduction already indicated: a set-theoretical reduction and a probabilistic reduction of variables. Each grade obtained by an individual i on a test j is considered as a random variable, and not as a function determined by its ideal score in terms of an underlying reference-frame. It is the distribution of this random variable which, obviously, is provided by the application of tests to the population. The determination of the reference-frame is then posed in terms analogous to those of Guttman's problem. The reference factors can be chosen according to an identical criterion. In the language of geometry, let n = original tests and N = individuals. In a space of n dimensions, the N points representing the individuals scored according to the tests are distributed in a dispersion, where the calculation produces privileged directions which serve as new and more natural axes and represent the factors or dimensions of the required reference frame. The structure constructed is thus linear in the mathematical sense, that is, intuitively speaking, the reference-frame is thought of as a bundle of axes, by relation to which the representative points described by the grades on the tests are assessed, and treated like the extremities of *vectors* issuing from the apex of the bundle. By means of the hypothesis of linearity,[14] the calculation allows the substitution, for an anarchic and super-abundant multiplicity of original tests, of a structured and 'economical' multiplicity of 'pure' tests, whose combination should suffice to characterize the individuals.

The analyses of latent structures and Guttman scales rest on the same idea: in both cases the object is always to relate a plurality of empirical responses, in which qualitative distinctions initially appear, to a 'latent structure'. The project of quantification, particularly evident at the beginning of the research, fades out in the course of the work, so that the quantities introduced, in the final analysis, play only an auxiliary role in regard to the structure put into effect. It is in this perspective that the following remarks of Guttman are to be understood (Guttman 1954, p. 250):

> In fact, the only psychologically clear problem that we posed about the metric was the question of the zero point. We did not really know what in the end we expected as a metric, but we only had the feeling that it would be nice if there had been 'something more' than a simple linear order . . .

Is not this something more – the interpretation of higher order components – the very structuring of the qualitative multiplicity?

5.16. Thus, the value and the significance of a measuring technique depends essentially on the fruitfulness of the mode of structuring that it involves, and this is what is meant by a scientific dialectization of quality. It might be thought, however, that the 'law' of the transition from quantity into quality, viewed from the epistemological point of view, introduces into science an inverse process radically opposed to the qualification of the quantitative. I have dealt with this point in the second part of my study of economic methodology (Granger 1955), where this emergence of qualitative differences between the micro-object and the macro-object of economics is considered necessary for the restructuring of the phenomenon. Here I shall limit myself to noting that in the most general sense quality is introduced into the sciences of man only in the conceptualized form of a structural typology. In the preceding sections I have shown, through examples from social psychology, how the attempt to reduce a qualitative multiplicity leads to the construction of underlying abstract models, whose articulation can account for the distribution of observed phenomena. In the case at hand the irreducibility of two or more levels of phenomena is established: individual or interindividual behavior studied by the psychologist, social movements and mechanisms described by the sociologist or economist. The qualitative distinction between these two modes of the human phenomenon seems to be bound to the quantitative transition from the 'small' to the 'large' group. Attempts to reduce the explanation of the macroscopic to a summation of microphenomena fail. It is still true, substantively speaking, that no doubt the object of science is the human being, necessarily individualized in his concrete existence. To describe the multiplicity of the aspects of this human being in a simple way, according to his involvement in groups of varying sizes, as a qualitative universal, is to resign oneself to fixing science in an infantile stage. But it is equally impossible, as the history of sociology or economics shows, to claim to reduce these differences to a simple quantitative variation of parameters, by assuming the homogeneity of the levels of the phenomenon. Once again, the way in which knowledge is acquired is through the reduction of the qualitative to the structural. In the 'large' group a different structure is postulated, containing types of relations which do not appear in the 'small' group. Thus it is not enough to simply amplify the structures of communication or hierarchization which permit the appropriate interpretation of the phenomena of the small group, in order to explain the macrophenomena of stratification and conflicts that arise at the level of a society; a new perspective is necessary. However, the social psychologist, whose object *par excellence* is the 'small group', often pretends to think that the applications drawn from his own studies should serve at the same time to solve problems posed at higher levels, and that, for example, a class conflict is only the magnified transposition of friction within a team . . . Since Marx and Keynes, political economy has almost universally renounced this illusion of the psychologist and has refrained from treating macro-factors and global magnitudes as the simple additive results of micro-factors and magnitudes on a smaller scale. From here on in one can predict that an analogous recognition will soon spread to all the disciplines which treat the human fact. The qualitative 'leap' which appears in the organization of phenomena, is thus expressed epistemologically in terms of a change of phenomenology, and consequently in terms of a conceptual instrument. And one can not say that it is only a matter of a transformation which is, so to speak, continuous with the primitive structures, linked to the growth of a determinate magnitude beyond a certain threshold: for beyond this threshold, the very nature of the measurement has changed. *It is no longer the same thing* that is being measured, for quantification comes about differently in the initial structure and in the new structure: 'national income', for example, is a homonym of individual incomes or corporation incomes; but it has a completely new functional meaning in the macrostructure. It corresponds to a technique of evaluation that is very different from, and is in fact no longer an *income* at all, in the microeconomic sense. Similarly, it must be admitted, for example, that a sentiment, such as that studied by the social psychologist, changes its sense profoundly when the sociologist makes it an element in a description of social life, since in the second case it is defined by a radically different structuring of the objective set [of subjects] considered.

Thus one of the most difficult problems posed by the modern development of science is that of the relation between phenomena which are juxtaposed or superimposed, the articulation of which must nevertheless be uncovered if we are to understand man. A homogeneous, qualitatively differentiated object, the progress of knowledge is gradually replaced by a hierarchy of structures, valid at different levels, without which the legitimate wish for an accord between these universes is still not fulfilled. We shall content ourselves, at this point in our analysis, with underlining the importance of the problem, while reserving for the last chapter the outline of the paths toward synthesis, whose shape, I believe, we can already see.

DISORDER AND ORDER

5.17. Under their diverse forms quality-quantity relations have appeared in the human sciences as those of the astructured to a structure, and these are relations which shatter the all too rigid concept of quantity, by requiring the mathematician to make an effort at invention whose first fruits we are only now still discovering. Thus far I have limited myself to presenting an analysis of certain examples of the method; it would be good to conclude our study with a more general examination of the techniques of structuring, which the human sciences, in the present state of their research, seem capable of employing.

The general problem which is posed is that of the transition from *disorder* to *order*, in the usual and broad sense of the word. It is not that disorder is the spontaneous category in which human facts must be presented. It can be thought on the contrary that the characteristic of the human phenomenon is that from the first, it is given as ordered and regular. Societies are stratified, conduct imitates certain types; and the object of the psychologist, and sociologist, are apparently offered as structured. Natural phenomena on the other hand, offer themselves to the eye of the naive observer under the species of a multiplicity without reason, where only patient analysis can unravel regularities and laws. In fact at the level of the *perceived* object, the human fact is always grasped as immediate meaning, or at least with the promise of meaning. The explanation and the interpretation of these meanings does not depend on science, but constitutes, on one hand, the practice of life, and, on the other, the task of the philosopher. The scientist, who aims at constructing *models* of phenomena, thus cannot confuse this order of meaning with the abstract schema that he claims to establish. We must reverse the sentence Merleau-Ponty employed to define the purpose of the phenomenologist: "It is a matter of describing, and not of explaining and analyzing." On the contrary, it is a matter of explaining and analyzing and not of describing, if description means the comprehension of meanings. Once this reduction is made, the human fact becomes an *object of science*; it is not that it is reduced to the simple dimensions of the thing, but its density of meaning is, as far as possible, preserved, neutralized, finally objectified. In these conditions, a philosophy, considered as hermeneutics, naturally keeps its place alongside science, whatever its state of progress, but it can only replace science by imposture, just as a similar imposture would suppress philosophy for the benefit of science.

So, far from grasping in the human fact a spontaneous order, the scientist finds himself in the presence of a disorder, that is, of a multiplicity of events whose constitution he must seek as completely organized. The answers to questionnaires cited above as an example, or automobile traffic at a red light, or housewives' choices among different products in a store, represent disorder. According to Bergson this is a wholly negative idea, and that is true. But the philosopher goes too far when, assuming two types of order, he defines disorder as the feeling of the absence of a certain expected order, and the presence of an order of the *other* type. The idea of disorder, he says, "would objectify, for the convenience of language, the disappointment of a mind that finds before it an order different from what it wants" (Bergson 1911, p. 222). On the level of a psychology of disorder, where above all Bergson seems to remain, the purely negative interpretation is not admissible: have not gestalt psychologists brought to light a positive perception – in a certain manner one already structured – of the 'jumble'? In any case, if on the phenomenological level the Bergsonian idea of disorder as privation appears justified, one cannot bind one's fate up with that of the theory of two orders, the mechanical and the vital. The order to which the view of disorder refers as its horizon remains open, possible, ambiguous. Moreover, in the view of the scientist, it is not a simple indeterminate absence, for it depends on techniques of observation and intervention which he has at his disposal. Insofar as his object is not merely a percept, the psychologist or the sociologist grasps it above all as disorder on the horizon of a possible structuring. It is the active analysis of the object which permits him to organize this qualitative 'jumble', either by specifying this promise of order, or by inventing a new type of structuring. It is thus important for the epistemologist to canvass, if possible, some of the schemes or motifs of structuring currently in use in the human sciences.

CLASSIFICATIONS

5.18. The first scheme that presents itself as a set of qualitatively differentiated elements is classification. In the sciences of man, the sociologist and the psychologist make constant use of it. However, a distribution into classes in the realm of the human fact never completely satisfies thought, since it always involves the artificial and the static. There are, moreover, different degrees of structuring and cohesion in the process of classification. At its lowest level, it consists in dividing one set into disjoint sub-sets defined more or less exactly, and without any synthetic motive that connects them, other than the requirement that they be exhaustive.[15]

A little more formally elaborate is dichotomous classification, and under an apparently more general form, the classification which might be called *Linnaean*, whose graphic representation is a tree-shaped pattern. I have already stressed above the importance of dichotomy to which it is always possible to reduce every tree-like classification, by means of the adjunction of supplementary subclasses which unite a group of classes in order to oppose it to a single class, distinguished at the same level. In this way, one artificially rediscovers the dichotomous schema.[16]

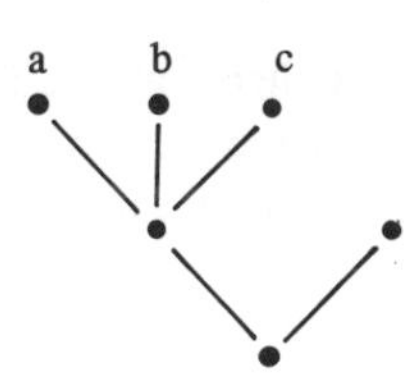

(1) Linnaean classification

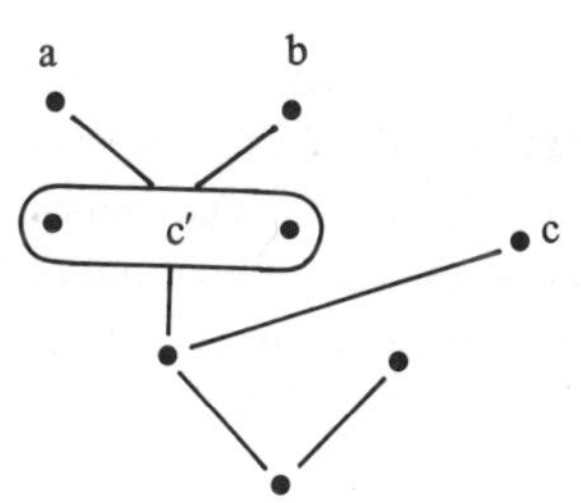

(2) dichotomization of 1. by adjunction of class c', with a supplementary level.

Thanks to this process, the qualitative multiplicity is *stratified* into levels, relatively arbitrary no doubt, and is distributed into ramified *lines*. The very nature of the bond between elements on the same line naturally depends only on the objective domain considered: it can be an association, an inclusion, a resemblance, indeed even a dynamic relation of influence. Linnaean classification is by virtue of its very abstraction, a type of structure that is rich in possibilities. However, it is in the direction of a more elaborate abstraction that the classificatory instrument is developed. Such is the case for the *combinatory* classification studied by Bachelard in the natural sciences under the name of multiple or 'crossed' order. Elements are distributed according to two or more dimensions, giving rise to a multiplication table, for example, such as that of Gurvitch's classification of types of sociability, or in the characters of Heymans-Le Senne. Such a structuration already assumes a higher degree of abstraction than that of Linnaean classification. It is not in fact the elements themselves which are distributed in classes, but two (or more) differentiated components of these elements; and this is a process which somehow isolates the purely ordinal basis of the Cartesian representation.

5.19. The notion of order *stricto sensu* thus appears to be totally detached from all numerical implication. It refers only to a certain mode of being of the elements of a whole in their relation to one another. An order is thus defined on a set of elements by a certain relation between these elements taken in pairs; depending on the properties of this relation, this order will be represented by a series of points on a line, on a circle, on a tree, or on a network. A mathematics of order sets forth the consequences of such an abstract determination, and thus furnishes the framework for a structuring. A more elaborate analysis of apparently quantitative concepts in the sciences of man has shown more than once that magnitude is here only a rather arbitrary garb, and that what solely exists, as a reasonable and motivated schematization of experience, is a structure of order. This is the case for Pareto's 'utility' in economics,[17] and with a great number of scoring sytems in psychological tests. An attempt to axiomatize can only bring to light the quantitative illusion, because it makes explicit its presuppositions with exactitude. I shall soon examine in general the conditions and difficulties of this enterprise.

LINEAR STRUCTURES, VECTORIAL SPACES

5.20. Reflection on classification has led us to the mathematical concept of ordered structure, raising here and now the problem of axiomatization. I should now like to say something about a second schema, essential to the structuring of a qualitative ensemble, examples of which have already been given. This is the schema of linearization. I shall try to outline its general significance here. Since the aim is not the development of a mathematics, but only the critical analysis of the treatment of quality, my object will be above all to distinguish this scheme from a scheme of quantification pure and simple, and to show the structural significance of quantification itself.

In the presence of a qualitative multiplicity, the aim of rational thought is not only to classify, that is to structure according to a relation of order, but if possible, to reduce the complexity of phenomena to certain components or dimensions, as Descartes said. Types of combinatory classifications already satisfy this requirement, but very incompletely. The more or less avowed ideal which orients the research worker is the construction of a model of the phenomenon in which the qualitative differences are decomposed into variations of intensity in different dimensions which are admitted as fundamental. Such a schema is immediately available in the domain of numerical magnitudes through the simple notion of homogeneous linear equations. The (quantified) phenomena X is analyzed as a sum of products ax_i, where x_i represents the 'unit' of a variable or fundamental dimension, and a the coefficient of 'intensity' of this variable:

$$X = ax_1 + bx_2 + \ldots + px_n.$$

It is thus under its quantitative and indeed numerical form[18] that the ideal of linearization is spontaneously aimed for. It thus assumes a possibility of a rough reduction of the qualitative phenomenon to a numerical magnitude. But this naive interpretation of the ideal of linearization is by no means essential. The modern conceptions of algebra enable us to dissociate different structural aspects of quantification within this scheme.

5.21. Simply consider, then, any abstract objects, and certain operations on these objects. We want every object X of the set to be *constructable* on the basis of certain privileged objects x_i, by means of two types of operations. One, which we shall call (metaphorically) additive, must be endowed with properties formally analogous to those of arithmetical addition; more precisely, it must determine, on a set of objects, the structure of an Abelian group. The other operation is more ambiguous. It assumes the existence of a second system of objects, perhaps distinct from the first, playing the role of 'coefficients', in the linear construction. Their properties must be formally identical to certain properties of integers, or eventually of rational numbers; that is, they must constitute a *ring* or a commutative field on which operations analogous to algebraic addition, multiplication, and division can be performed.[19] The linear schematization thus assumes the existence of two sorts of entities: *objects* forming a group, and *operators* forming a ring or *field*. Neither is necessarily thought of as numbers; but even when the field of operators is a set of numbers (as happens in factor analysis of aptitudes, for example), it does not follow that the set of objects themselves is strictly speaking quantified.

The mathematician thus demonstrates that this set of objects possessing the structure of a 'vectorial space', has the fundamental property which satisfies the originally formulated desire: in this set a family of elements chosen as a *base* can always be found, such that any object can be constructed from them by a linear homogeneous combination.

The fine Eudoxian theory of relations, in Book V of Euclid, already exhibits this duality of the domain of magnitudes and the domain of operators:

The notion of *measurement* includes the multiplication of a magnitude by an operator which is a number, and originally an integer (see definition 1 and 2). A magnitude is a part of another magnitude larger than it, "when the smallest measures the largest": *ὅταν καταμετρῇ τὸ μεῖζον* [when it measures the largest]. "To be measured by" here means being equal to a *multiple of* – *πολλαπλάσιον* – [many times as many] obtained by the application of an integer-valued operator.

Definition 4 includes the axiom later attributed to Archimedes: "Two magnitudes bear a relation between them when it is possible to find a multiple of one which exceeds the other" and this involves essentially the notion of operator.

With respect to the first abstract conception of magnitude in general, it is no doubt to be found in Grassmann's *Ausdehnungslehre* (1862), which introduces explicitly what Bourbaki calls vectorial space, and demonstrates general theorems proper to this sort of structure, in particular the property just dealt with, a representation by means of a basis of *n* elements.

5.22. If we attempt to analyze the quantification, strictly speaking, of a set of abstract elements, we see that in addition it requires more.

(1) The quantified elements are capable of a total and simple *order* like that of the points on a line (and not on a circle, or a network).

(2) They can be combined with each other by an associative and formally symmetrical operation analogous to arithmetical addition, with its inverse being subtraction (not everywhere defined, however): they belong to an algebraic structure of *monoid* (cf. Bourbaki 1974, vol. I., chap. 3). A relation of equality must naturally be defined for these elements, among which there must still be introduced the neutral element o, such that, for any element x, $x + o = x$.

Moreover, they must be combinable with whole numbers by an external operation analogous to arithmetical multiplication: one must be able to speak of a quantity n times greater than another.

The quantified elements thus constitute (at least if the subtraction is everywhere defined) a "module" on the ring of integers, or ultimately a vectorial space on the field of the rationals or the reals.

(3) The order must be compatible with the algebraic operation of addition, in the following sense: if $a > o$, for all x, $x + a > x$.

As we can see, this analysis applies to quantity in general. The notion of 'continuous' quantity obviously assumes the adjunction of a fourth type of requirement, which is suitable for guaranteeing meaning for the expressions: "quantities approximately equal", "converging series of quantities", etc. But the decisive property of the notion of quantity seems to us to be attached to the third condition, which binds a structured order to an algebraic structure: and nothing like it has yet appeared in our scheme of linearization.

5.23. The nature of the process of structuring on which the linear schemata depend is now apparent: quantities figure in it only as auxiliaries, and naive thinking is tempted to identify the elements of the vectorial space with magnitudes, by implicitly endowing them with predicates tacitly recognized by [their] intuitive quantities. (Such as measures of segments on a line, for example.) It is thus that, all too often, one assumes, without discussion, a topology, properties of proximity which are not at all included by the strictly algebraic definition of a vectorial space. The progress of the method will thus necessarily include an increasingly conscious and refined analysis of presuppositions, and of their value of representation in relation to phenomena, in short, their value in an axiomatic elaboration. It seems permissible to assume that in the majority of cases, attempts of this nature can only be local, tentative and fluctuating. At the limit of research, it is still necessary to uncover them. Moreover, it would be imprudent to assert that the type of structure analyzed above should remain preponderant in the sciences of man. Certainly it corresponds to an essential stage of general algebraic thought, and appears to represent by virtue of this a point of departure for the construction of models. But the mathematical imagination can no doubt conceive of eventually providing combinatory developments of another style. Up till now, all the formal tools of the natural sciences have rested on linear schemes: classical algebra, infinitesimal calculus,[20] theory of probability. It will thus not be surprising that attempts at schematization in the sciences of man will begin right away on this terrain which is not only fertile but already well known. However, knowing the difficulties to be encountered by such a representation of phenomena, one cannot conclude that science is impossible here, but only that the tools which it currently employs are inadequate.

THE RANDOM SCHEMATA

5.24. Having thus attempted to outline in a radical way one of the most general schemes of this dialectization of quality, I should like to evoke a final motif of structuring, whose role has been emphasized several times: the recourse to random schemata. It will not escape the reader that in our recording of schemata in a particular order: classification, linearization, recourse to randomness, I have not made any claim to establishing a hierarchy. In fact, the three motifs described here are found on different levels and their interrelations are obvious. From classification we have been led to vector spaces, as from the static to the combinatory; probability schemata, with respect to their substructure, again depend on linear thought: they introduce vectorial spaces. But they clearly offer traits of organization particularly favorable to a representation of human phenomena: they conceptualize in a quite natural fashion the '*almost*', the '*sometimes*', the ὡς ἐπι τὸ πολύ [for the most part]. I have no intention of summarizing a critique of the usage of probability in the sciences of man.[21] I would simply like to formulate some reflections on the role of random schemas in the structuring of the qualitative.

If the qualitative aspect of the understanding of the human fact appears from the perspective of science as the as-yet-unstructured, it is quite natural to apply to it the concepts of probability, which aim explicitly at the structuring of a certain disorder. In fact the notion of randomness itself was originally arranged for an objectivization of human behavior. It began, as is well known, with the speculations of Fermat and Pascal on the redistribution of stakes between players in a suspended game, and a half century later, in the celebrated writings of Jacques Bernoulli, it also constituted a theory of 'chances' – i.e. a theory of reasonable decision making under conditions of uncertainty – and a theory of global predictions of the probable. For more than a century the other point of view prevailed, and was made explicit on the philosophical level in a frequentist conception of probability, and on the technical plane in a theory of probability distributions, facilitated by the extensive application of infinitesimal calculus. Although the former doctrine continued to be represented by somewhat dissident interpreters, or to be marginally expounded by the defenders of 'objective' probability, it did not really attract attention until the recent period, when the problems of probability have been presented anew, as problems of *decision*, in particular in *game theory*. Independent of attempts to provide an axiomatic basis for the notion of probability – which naturally reflect this duality of perspective – it is indispensable to take account of such an opposition of points of view in order to grasp its application to the qualitative aspect of human facts.

5.25. The point of view of probability distributions essentially involves the following: given an event susceptible of various outcomes, one must coordinate the set of actual realizations of these outcomes with an *invariant* and exhaustive scheme of *potential* realizations. The sets of actual realizations can be called 'samples', and the schemata of potentialities will be a law of distribution of frequencies. The most familiar case would be without doubt the game of heads or tails, where the event is capable of two outcomes. A series of n trials constitutes a sample in which the two outcomes are distributed 'at random': that is, a determinate series, taken in itself and isolated, cannot be predicted with absolute certainty. The law of distribution which corresponds to it is, however, perfectly defined; we know that it is obtained by exhaustively enumerating all the *a priori* reasonable samples. The 'probability' of a determinate composition in a sample having two outcomes (p heads and $n - p$ tails) will be defined as the relation of the number of possible samples realizing this composition to the total number of possible samples. The character of isolation and qualitative differentiation of a given series is thus dialectized by its insertion in the order of possible frequencies of the schema of distribution. One can state once more that the process of structuration *does not have a direct bearing* on the qualitative given: the recital of events taken in itself remains what it is on the level of qualitative perception; the structuring performs a *transmutation* of the immediate object into a mediate, potential, organized, and finally, scientific object.

The example of heads or tails may seem too far removed from the domain of the human sciences. If one prefers one can substitute for it the more complex example of errors of measurement. A single subject making the same measurement in a series of *n* times acquires *n* results, of which some at least are no doubt different from one another. These differences can be considered as variations of the same phenomenon. The theory of errors attributes probability to each outcome.[22] Thus an abstract structure of potentialities is substituted for qualitative variation, considered as an event. Since Bernoulli, Laplace, and Poisson, the multiplicity of conceivable schemes has been restricted to a small number, corresponding to particular, rather natural hypotheses, which have been uniformly exploited in the natural and social sciences: polynomial distributions (like that of heads or tails), the Laplace-Gauss distribution (like that of errors), the Poisson distribution (for events of which each outcome is 'rare', and becomes increasingly rare as the volume of the sample grows).

The typology of chance thus appears rather impoverished, as soon as one gives substance, so to speak, to the notion of distribution. Such a simplification assuredly depends on a bias toward restrictive hypotheses at the beginning of the construction of schemas. But it also depends, it must be admitted, on the processes of approximation imposed by the conduct of the calculations. This ascetic exercise was apparently necessary in the construction of the mathematics for randomness. Whatever the case may be, the structural reduction of the qualitative recital of events by means of random distributions is strict. The sciences of man, while retaining its benefits, are not slow to rid themselves of it.

5.26. It is precisely the second point of view which, by introducing chance as an element of a reasonable decision, in a certain sense makes the probabilistic schema more supple. In the types of random distributions, probability appears as a static feature of an object. With the theory of games it is introduced as an instrument of a decision. The conception of games of strategy – more precisely, of the simple duel – involves the establishment of a table of players' gains and losses, as a function of different tactics open to him and to his adversary. One notices first that an initial reduction of the qualitative is at work in this schematization of behaviors, which are related to a discontinuous field of possible tactics. The elaboration of the model can be pursued by further assuming that the behavior of the adversary is already known as random structure. Every one of his tactics is used with a determinate probability. We have here recourse to probability as objective distribution, and we shall see later which norm of action can then be defined.[23] But one might think that such a hypothesis is, in most cases, deprived of empirical meaning, or at least that there is no conceivable process for estimating these probabilities. One thus attempts to describe a norm of decision on the basis of a simple table of losses and gains. The analysis then leads to the recognition that, generally speaking, no simple tactic satisfies the proposed norm, but that a certain new type of decision process always satisfies it. The new type of decision consists in the subject *randomly* choosing among the tactics open to him, each of them being assigned a probability calculated on the basis of the matrix of gains.[24] Probability thus reappears here no longer as the *constitutive principle of a structure of the object, but as a regulative principle of a structure of behavior.* The player's unique and originally qualitative act of decision appears henceforth to be integrated into a schema whose structure reveals order, although paradoxically it maintains the element of unpredictability.

CONCLUSION: DIALECTIC OF QUALITY AND AXIOMATIZATION

5.27. At the end of this analysis, it appears that quality, as it is presented in the social sciences, is disorder: a disorder given meaning, it is true, on the plane of perceptual comprehension. But just as the object of physics is revealed only when it is understood that a systematic and invariant interpretation of phenomena is not based on perception, so the human object is discovered as *object* only when one agrees to aim at disorder through this immediate sense, a disorder which recovers the unity of meaning on another plane, that of the phenomenon structured by science. The treatment of the qualitative in the sciences of man is nothing other than the development of this structuring.

It is commonly believed that the activity of structuring, which certainly requires the employment of methods of rigorous thought, is a quantification pure and simple, and that in a narrow sense, there can only be a science of what is measurable. We have seen that it is nothing of the kind. If quantification appears as one of the most satisfying terms to which this step can lead in the most favorable cases, it cannot be the only modality that must be successful. In a much more general manner, it must be said that the qualitative is conceptualized by reduction of *isolated* differences to differences *integrated* in a coherent system of oppositions. This reduction lies at the very origin of mathematical thought, and in this sense one can recognize that this treatment of quality is a mathematization; for all that it is not an introduction of the *quantum*, and several adversaries of a rigorous approach in the human sciences, in attacking quantification alone, have been merely attacking clouds. (Cf. the suggestive study by John G. Kemeny, 'Mathematics without Numbers' (1959).)

This mediation of quality, which leads to the concept, is not a speculative mediation. In the arts of word and spectacle such a mediation, which leads to meaningful images that are more vigorously meaningful than the phenomena themselves directly perceived, will be demanded. But this 'understanding' of man does not have the characteristics required by science, and can only be integrated into a contemplative attitude of esthetic enjoyment or into a daily empirical practice which is itself a sort of art. A science of man can naturally replace neither the fine arts, nor the concrete individual practice of human relations. No more than physics, or chemistry can be substituted for the 'flavor' of sounds, smells, colors, or for the art of cooking.

Scientific mediation aims at another goal. It turns its back on the direct perception of meanings which most often orients daily empirical practice, but this is to prepare a model of phenomena which will more effectively frame a concerted, organized, rational practice. The process of structuring itself is found to be bound to this level of practice: the concepts which it constructs arise as strategic operators, and not as contemplative explanations. They are conceived in the context of a practice which puts them to the test and requires their incessant revision. It is this which brings forth quality and difference anew and calls forth a reformulation of structuring.

If this is the dialectic of the qualitative, then the attempts at axiomatic reduction, which begin to come to light in the sciences of man, appear from here on not as an end towards which empirical understanding tends, but as the movement which makes this renewal possible. To axiomatize means to transpose a latent structure into an adequate language, and to give a form of provisional equilibrium to concepts. The contradictions engendered by practice manifest themselves all the more distinctly when the structures have been more explicitly thematized, and objectivized in an axiomatic system. So that, far from being a factor of academic immobilization of knowledge, the axiomatizing tendency must be recognized more and more clearly as one of the moving forces of a dialectic.

CHAPTER VI

STRUCTURING AND AXIOMATIZING

6.1. The treatment of quality as it has just been presented can be considered as a paradigm of the construction of concepts. The treatment rests on the building of explicitly structured abstract models. Now, if one examines this work of formal thought under a more general aspect, one is led to outline the traits of its technology, and to look for the direction of its movement which seems to lead structuration toward axiomatics. This is a movement too often misunderstood as the result of a misplaced, hyperbolic and vain effort to reduce human facts to a pure play of thought. In this chapter I hope to show first, with the aid of a technological study of models, the complementary character of the two undertakings in question: the structuring of the phenomenon and the axiomatization of structures. We should then be in a position to state precisely the epistemological significance of axiomatic systems in the sciences of man. It is a significance profoundly different from that which they have in geometry, although a prejudicial misunderstanding can arise and develop in this regard. After attempting to dissipate this ambiguity in the notion of axiomatic systems, I shall then examine more closely one of the attempts to apply the axiomatic method to the elaboration of a determinate concept for the sciences of man: that of 'rational' behavior.

6.2. First I want to give a general idea of the manner in which mathematical formulation is distinguished from the 'vulgar' formulation of a model. Simon (1957) provides a rather instructive formulation, in which he attempts to reduce to a system of equations, a psycho-sociological model presented without mathematical concepts by Homans (1950).* We can limit ourselves in this preliminary examination to outlining, as an example, the points on which the attempt at transposition has an essential bearing.

(1) First of all it is apparent that Simon takes great care to dissociate *metric hypotheses from weaker hypotheses,* like those of order, for the definition of variables. Ordinary language generally leaves this distinction vague and so acts as a brake on the dialectization of quality to which we devoted the preceding chapter.

(2) In the mathematical model, the necessity of treating certain variables as statistical means is formally emphasized, and precise hypotheses are

eventually advanced to deal with the behavior of magnitudes which fluctuate around these means.

Here again, ordinary language masks the unevenness between individual variables and average variables, and blocks the dialectization of 'chance'.

(3) The relations between the variables, which the nonmathematical formulation expresses vaguely as one of direct or inverse dependence, are made precise by the mathematician. In particular, *linear* hypotheses are explicitly introduced, when they tacitly constitute the common foundation of all non-formal accounts. Thus the simplifications and reductions from which deductive thought proceeds are readily brought to light.

(4) Finally, mathematical formulation requires the statement of precise hypotheses on the behavior of variables *in the neighborhood of equilibrium*. The very idea of equilibrium and stability remains simple and confused in intuitive thought. The mathematical treatment enriches it, at the cost of hypotheses which often are difficult and delicate to formulate. Thus, it is always a matter of the dialectization of the intuitive, moving from a passive understanding closed on itself to an active, combinatory and open one.

Other examples of the same sort, easily discoverable in the works of economists, and certain psychologists, would testify to the same orientation. The mathematical elaboration of a model thus becomes indispensable from the moment that a rather refined and rigorous analysis of notions comes to require precise choices between hypotheses. All real structuring tends towards this mathematization.

'ENERGETIC' MODELS AND 'CYBERNETIC' MODELS [1]

6.3. The example just examined showed the role of a mathematical formulation of models. It does not, however, offer the only reliable testimony. For the relations of variables that it uses are somehow related to a *system of forces*. They describe a homogeneous machine in which everything is situated on the same level. Although this kind of model can still be used to describe many phenomena, in present-day science it is nevertheless opposed to another type, rich in promise, which introduces relations between variables on two levels. On top of the relatively rough interplay of *energies* it superimposes the relatively finer interplay of *information*. Plainly, it is not a matter of reintroducing the experienced notions of desire, command, information transmitted and understood in a model. The models of the second type remain abstract, indeed they formulate these notions which are themselves treated as abstract relations, functionally irreducible to energetic relations, even though their support is of the same nature as that of the energies whose interplay they regulate. A two-level model for human behavior does not presuppose any dualistic ontology at all: the levels are two types of structuration which combine with one another, and are by no means two kinds of being.

In a previous work on the methods of political economy I tried to provide a classification of models used by this science, according to the degree of conceptualization of the time variable. One would doubtless find beneath this finer classification the rougher but more general one proposed here.[2] But what is possible to try in economics is not yet possible for the sciences of man as a whole, as these sciences have not yet reached the same stage of conceptual elaboration. In other respects this duality of models gives rise to an essential advancement of science. It is worth noting that such a distinction can already be met in a distant precursor of the structural treatment of human phenomena. When Condorcet, in his *Essai sur l'application de l'analyse à la probabilité des décisions rendues à la pluralité des voix* (1785), wanted to construct abstract models of an electoral body, he postulated first a homogeneous decision mechanism, in which

> all those who give their vote have an equal wisdom, an equal accuracy of mind, of which they all make equal use, [that they] are animated by an equal spirit of justice, and finally [that] each of them has voted following [his own mind] as it would happen if each one gave his opinion separately, or what comes to the same thing, that in the discussion each of them has had an equal effect on the opinions of the others (p. 3).

The probabilistic determination of the outcome of the voting is a result of the composition of the characteristics of each voter. The organizational scheme of the model is thus comparable to one of a machine transforming in-put flows of energy from out-put flows, in accordance with certain laws without the intervention of regulation of another order. The same would not be true of models of the 'second sort' (cf. Granger 1956, p. 103 and 113) for which Condorcet introduces asymmetrical reciprocal influences among the voters, and an evolution of individual opinions in the course of successive votes. In this case, on top of the direct determination of the result by the primary interplay of individual voting probabilities, there is superimposed the interplay of influences, as if a secondary, regulating circuit was added to the principal energy transformation circuit in a machine. Of course, Condorcet did not make the concept of heterogeneity explicit in the models of the second sort, and moveover the mathematical apparatus which he used would hardly permit its exploitation. This distinction, however, prefigured the duality that can be recognized today as a leading idea for the classification of models. I shall show this with an example.

6.4. The phenomenon to be structured will here be a linguistic one, apparently discovered by Zipf (1935). If the words in a sufficiently long text are classified by their frequency, we notice that there is a simple relation between the number of words of equal frequency and the value of this common frequency. The simplest approximate law, valid in an average zone, is the following: The number of words having a frequency x is close to k/x^2, k being a constant which depends on the text (or the language) considered; or: the frequency of a word is inversely proportional to its rank. The first step in structuring involves improving this rather loose formula, by considering the distribution of words in a sufficiently long text as a statistical result. Conforming to the process already described, the event – a text or collection of texts having meaning and carrying a message – is reduced by the neutralization of its content. It is treated as a mere representative sample of the linguistic substrate which constitutes its matter, and this reduction is legitimated *a posteriori* by the invariant character of the distribution. At the 'microscopic' level, the distribution of words in a sequence obviously depends on the meanings which the writer wished to send. But at a macroscopic level all this happens as if laws of a different sort determine the distribution of words, independent of meanings. This is a new example of the essential pluralism of epistemological levels. The problem is then: (1) finding a law of statistical distribution which fits the data more closely; (2) constructing a model of random selection such that a sufficiently long series of operations leads to a *stationary* distribution of this type. The phenomenon is finally 'explained' as a figure of equilibrium of a homogeneous process, consisting here in a series of random choices, bound by as yet undetermined conditions.[2]

Simon achieves the construction of a more approximate law of distribution in two stages in *Models of Man*. First of all he proposes an empirical formula: $f(x) = a(b^x/x^k)$, and he indicates some conditions of its application to the phenomenon.[3] He then proposes a new formula, derived from the Euler function B, which, while less simple than the empirical formula, sufficiently fulfills the required conditions, and has been integrated beforehand into a well-known and analyzed mathematical structure. The aim of this formal elaboration is, naturally, to facilitate the deduction of the resulting distribution on the basis of the hypotheses of the model. The second stage of the construction consists precisely in stating the postulates which define this model. I reproduce them below, to show their character:

(1) The probability that the $k + 1^{\text{th}}$ word of a text has already been used x times is proportional to $x \cdot f(x, k)$, where the function $f(x, k)$ is the number of different words each of which appears x times in the series of the first k words of the text.

This postulate thus expresses a certain inertia in the usage of a language, its propensity to self-imitation (factor x), and the limitation of its effective vocabulary (factor f). The same is true of the second postulate:

(2) The probability that the $k + 1^{\text{th}}$ word is a new word is constant. The mathematical analysis applied to these structural postulates furnishes equations with finite differences which determine stationary solution $f^*(x)$ that no longer depends on k when it is large enough, and this solution has the desired form.

Thus, the structure of a model is defined which would justify the remarkably invariant distribution of the utilization of the words in a dictionary. The state of equilibrium which determines this stability results not from a superadded process of regulation, but from the very conditions described by the model. This would still be the case in a Walrasian system of economic equilibrium, or in the models of the first sort that Condorcet established for electoral bodies. Such a structuring can be labelled positivist, to the extent that it somehow remains at the surface of phenomena. The hypotheses on which it is based concern only the abstract form of the sequences of events, and no doubt there is a very wise bias for this at a still uncertain period of science. In this way analogies are brought to light whose fruitfulness could later prove decisive. Simon's distribution can be applied, for example, to many apparently disparate phenomena (cf. note 3, p. 199); but this uniformity of appearance poses a problem rather than resolves one. The concern with a less formal interpretation manifests itself moreover in our author who devotes an important part of his chapter to commenting on his models by introducing the intuitive notions of *association* and *imitation*. In a novel like Joyce's *Ulysses*, says Simon, it is clear that the frequency of a word like the hero's name is determined differently from that of the plural pronoun. In general, the reappearance of a word in a text will be related to two distinct phenomena: one of contagion or of internal 'association', and the other of 'imitation' of external paradigms common to all users of the language. The law of sequential probability is interpreted from this point of view as the resultant of two tendencies. But there is nothing really new with respect of the first, strictly positivist, version of the model, and we have not left the domain of homogeneous structures.

6.5. Now, there is another formulation of the same phenomenon, which will allow us to illustrate the contrast I have noted. Benoît Mandelbrot in his 1953 thesis (*Contribution à la theorie mathématique des communications*; cf. also his (1954)) considers, among other things, the properties of language considered as an instrument of communication. Following Shannon's ideas, he defines communication as a game of strategy in which the sender and the receiver are allied against 'nature', a source of 'noise'. Given a repertoire of elementary signals whose transmission assumes costs and a set – limited or practically infinite – of 'words' or units of meanings, the problem is then to determine the rules of coding these words in the most economical and certain way. If there are as many signs as there are 'words', it would obviously be sufficient to class the first in order of increasing cost, and to attribute to them as meanings words taken in order of decreasing frequency. Otherwise it is necessary to construct a rule of coding employing several signs for each word, but which nevertheless minimizes the average cost of a message, taking account of the frequencies or probabilities of employment of each word, and of the quantity of information which the message must transmit on the average. Such a coding constitutes a strategy in the sense of the theory of games, and Zipf's distribution should appear as a consequence of an optimal solution of this communication problem. It is clear that the model here proposed takes account not only of the lexicological material, but also of the informational function which it assumes in the system of the linguistic fact.

Let us schematize Mandelbrot's approach more precisely. The first difficulty derives from the concept of cost of transmission, in which we risk the chance of being lost in the as yet unexplored region of a refined psychology of communication.[4] Fortunately, it happens that by means of reasonable hypotheses of approximation, one can demonstrate that the cost of the nth group of signs classed in order of increasing cost, where n is large enough, takes a simple form, and does not depend on the individual costs but only on the total cost and is expressed as a function of rank. The result is that a criterion of economy, at constant information, leads to a coding strategy itself expressed in accordance with a simple law, where the variables of cost have disappeared:

$$p_{(x)} = P \cdot (x-m)^{-B}$$

This formula, in which $p_{(x)}$ is the probability – or frequency – of the word of rank x, obviously reduces to Zipf's simple formula when m is negligible and $B = 1$.

Mandelbrot's theory thus involves envisaging language, at a *certain level of structuration*, as a solution of a specific problem, a solution which does not correspond to the *resultant* of a system of forces or of some abstract determinations, but to the optimal *strategy* in a communications game. This game consists in coding information by pulling out groups of signs corresponding to words. A norm of action is posed: minimize the average cost of the operation. It is as if real languages, to the extent that counts have been taken, responded to optimal solutions: the significance of the 'natural' character of the *word* as a semantic unit would then be given in terms of this property of optimum. It is worth stressing here that the structuring thus introduced is placed at a determined level, without in any way aiming at exhausting the properties of the linguistic object. It is in this sense that Mandelbrot can speak of a 'thermodynamics' of language, for thermodynamic models play a completely analogous role in regard to the phenomena studied by the physicist. Neither the phonemic infrastructure nor the syntactic and phraseological superstructure are thus envisaged. It is quite obvious that Zipf's law does not at all prejudge the freedom of expression of the language user; the constancy of lexicological frequencies plays only the role of a substratum. Similarly, Mandelbrot's model in no way competes with the structures described by the phonologist.

But the very formula of the 'canonical' law stated above suggests another remark. The law involves three parameters, one of which, B, is interpreted by Mandelbrot as characterizing, within the limits of canonicity, different types of language use. By virtue of the thermodynamic analogies, he calls the number $1/B$ 'informational temperature': a high 'temperature' corresponds to an equilibrated utilization of vocabulary, in which rare words themselves are relatively well represented; a low 'temperature' corresponds, on the other hand, to a concentration of the common vocabulary on the words most frequently used. This latter case would be by far the most general.

From this we see that the model shows up new dimensions of the phenomenon. Far from being reduced to a pure and simple transcription of immediate empirical relations, it tends to establish a rather autonomous structure in order to raise to a higher level of abstraction the elements for which an indirect phenomenal interpretation must be found. The superiority of this type of model over the preceding one, insofar as it concerns human facts, is obvious.

6.6. We have already encountered this sort of model in Chapter IV, when discussing theories of learning as a dynamic game (4.22 sq.). Its most general characteristic is not, however, that it appeals to the schema of a 'game', for other less specific schematizations are no doubt possible. But it is true that the concept of game appears to be the first formalized realization of a structure at two levels. Borrowing the vocabulary of machines we may describe these two levels as the level of 'energy flows' and the level of 'information flows'. This language is no doubt too imagistic, but serves to suggest a fundamental division of systems for the determination of human phenomena. With regard to the intuition of experience, this hierarchization corresponds rather well to the opposition of substratum and sense, but we should avoid postulating a term for term parallelism between the structural elements of the information cycle and the experienced elements of meaning. In a model of this type, the second structural stage is just as abstract, just as formal as the first, and its treatment depends equally on a mathematics. Nevertheless the model thus constituted plays the part of a norm or a canon with respect to the phenomenon, in that it corresponds to a successful regulation. The case of the statistical structure of language, a very extended collective phenomenon, is obviously one of those for which a canonical distribution approaches closest to a law of equilibrium. Thus Mandelbrot's work represents a 'physics' of language, and begins, conversely, with an account of the physical problem as a theory of processes which permit the extraction of information from phenomena. All classical thermodynamics appears thus as a set of limitations imposed by nature on our strategies of induction. But to the extent the envisaged human fact becomes more specific, the canonical character of the results of the analysis diverges more and more from that of the laws of nature.

CAUSALITY IN THE MODELS

6.7. The duality of structure which thus seems to characterize models suggests a new position of the problem of causality. Certainly we are not dealing with a metaphysical problem; it is a matter of making precise the nature of the epistemological presuppositions governing a formulation of human facts. If it is recognized that a structure of the 'informational' type can be superimposed on a structure of the 'energetic' type, it must be asked to what extent the mode of relation which constitutes the second type reproduces that of the first type, and whether it is not, in fact, of a different order altogether. This is a problem which depends essentially on a technology of concepts, but which also involves the categorical nature of the object of the sciences of man. Let us pause for a moment at this point. In a homogeneous model, comparable to those of traditional physics, the relation between elements is expressed in functional correspondences. Clearly this term must be given a broad meaning which includes the stochastic relation – appearing for example, in Simon's model in Section 4 – as well as the strict relations postulated in the model treated in Section 2. Nevertheless, the essential historicity of human facts obliges us to stress the asymmetries which classical physics – save in thermodynamics – can ignore most of the time. The notion of determination must therefore allow for the possibility of a hierarchization of 'causes', which the functional relations only allow to be introduced accidentally. It is interesting to see, in this respect, the efforts made by the author of *Models of Man* (in Chapters 1, 2, and 3) to give a status and a precise definition for this idea of cause, in a perspective close to that of Carnap and the neopositivists. Simon presupposes, as elements of every empirical science, 'atomic sentences', irreducible units of the description of an object, and goes on to define causality as a type of relation between certain sentences. He introduces first of all 'state descriptions', sentences composed of the conjunction of all the atomic sentences (or, naturally, their negations) describing empirically the object of a theory. An 'empirical law' is stated as a molecular sentence derived from atomic sentences and consequently including a certain set of possible state descriptions. On the other hand, a set of "empirical laws" *determines* an atomic state, if all the state descriptions compatible with these laws include the sentence corresponding to this state (or its negation). Then, one can say that an atomic state a_i is *causally prior to* a state a_k if the smallest set of laws which determine a_i is included in the smallest set of laws that determine a_k. In this manner one infers a partial order on the set of atomic states, or more exactly, the phrases which describe them, and the causal hierarchy takes on a sense in extension similar to that of binary relations in the propositional calculus. This formal reduction is not by itself sufficient, for it is easy to see that transformations permitted in the propositional calculus can essentially change the causal hierarchy defined in extension without altering the logical value of the system of sentences.

Consider for example the two atomic sentences a and b, and the two empirical laws: $A = a$ and $B = a \rightarrow b$. It is evident that the law A is the minimal determination of a, and the set of laws (A, B) is the minimal determination of b. As the first set reduced to (A) is a part of the second (A, B), one might say that the atomic a causally precedes the atomic b. But if one substitutes for the two laws above, the laws $A' = b$ and $B' = a \leftrightarrows b$, the set of which is equivalent to the first in the sense of the propositional calculus, the same reasoning leads to positing that (A') is the minimal determination of b, (A', B') is the minimal determination of a; thus it must be said that b causally precedes a ...[5] In order to insure the invariance of the causal hierarchy, Simon introduces into his system a position variable, t (which could be time) and distinguishes between two sorts of atomic sentences: some, $a_i(t)$ will be 'observations', the others, $A_k(t)$, 'conditions'. An empirical law may be formulated by the schema:

$$(t)A_{i(t)} \rightarrow f(a_{1(t)}, a_{2(t)} \ldots a_{n(t)})$$

f being a complex sentence.

The result of these modifications is that two sets of laws determining the same state are not in general extensionally equivalent. Thus from the same empirical state one will not be able to deduce several causal hierarchies, except in the hypothesis of strictly different laws. However, the distinction between 'observation' and 'condition' does not appear capable of a satisfactory formal definition, if we remain only in the perspective of the object. It is necessary to introduce an action into the schema, and to define as a condition any state involving behavior of the experimenter. In other words, it is necessary to distinguish the *strategic* variables in the model; this distinction is certainly relative to an aspect of technique, a phase of human history. But we know that models of science – and particularly in the sciences of man – outline a provisional structure of the object.

This very requirement of asymmetry of the variables manifests itself in the application of Simon's ideas to particularly simple models. It is a matter of hierarchizing the variables of a system of linear equations. The position of the null coefficients in the matrix of the system then plays a fundamental role, that is, in the matrix of the variables whose influence is eliminated in certain cases. A hierarchization can be defined by means of two restrictions on the system as a whole. The first ensures that no subsystem of equations is overdetermined, i.e., no subsystem of laws is contradictory. The second ensures that every indeterminate subset becomes determinate once a suitable number of variables is fixed. By these means, it is demonstrated that there exists a partition of the system into subsets that are minimally determinate, plus, possibly, a remainder. By carrying the values of the variables determined by the subsets in the remainder, a new structure is obtained which can be divided in the same way, until the entire remainder has disappeared. This series of operations thus induces a partial order on the set of variables (*cf.* the table which illustrates the method, below). Of course, the equations themselves are not *observable* data; they are inferred from the sets of values taken on by variables in the course of experience, and the distinction of minimally determined sub-sets is thus equivalent to the recognition of strategic variables. But this distinction of privileged variables in reality makes no sense unless the simple homogeneous model is implicitly embedded in a complex model, which includes a second structural stage introducing human *action*. In fact, the only purely homogeneous models are those of an apparently speculative rational mechanics, for to the extent that our understanding of nature is developed, strategic discriminations, as in quantum physics, appear where the limits of the power of intervention are introduced. Thus the object 'nature' gives way more and more to the object of the 'technical complex' which is a prelude to a radically new extension of the scientific method.

In a homogeneous model it can be seen that analysis leads to a hierarchization of variables, resulting from a formulation of the notion of 'cause', but referring to the conditions of intervention of the experimenter, and suggesting in consequence the inadequacy of this primitive model.

Structure with six unknowns		matrix					
$A(x,y)$		a_{11}	a_{12}	0	0	0	0
$B(x,y)$		a_{21}	a_{22}	0	0	0	0
$C(z)$	linear equations,	0	0	a_{33}	0	0	0
$D(x,z,u)$	for example, for $B(x,y)$,	a_{41}	0	a_{43}	a_{44}	0	0
$E(u,v)$	$a_{21}x + a_{22}y = b_2$	0	0	0	a_{54}	a_{55}	0
$F(u,w)$		0	0	0	a_{64}	0	a_{66}

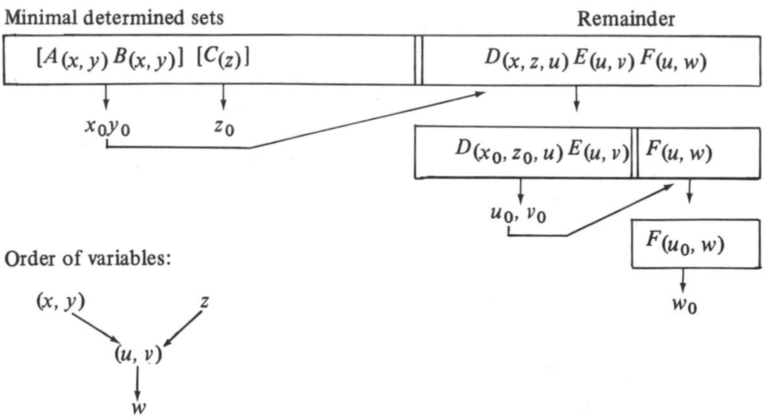

'CAUSAL' HIERARCHY OF VARIABLES OF A LINEAR SYSTEM

6.8. Such models are not in fact essentially distinguished from those employed in the sciences of nature. On the contrary, in the heterogeneous models we have presented, the relations of another order are superimposed on the first. In a phenomenology of the 'technical complex' there is a cleavage between a homogeneous infrastructure, a model of phenomena at the energetic level, and a superstructure articulated on the first. Clearly there is no question here of a genetic analysis, or a hierarchization of modes of reality. It is the double structural system taken as a whole which constitutes the object; what is described in the model at the informational level corresponds no more, no less, to the real than what is described on the energetic level. One can no longer speak of a 'causal' privilege of one of these aspects; the determination of the object, as soon as we can provide a model of it of the complex type, can be validly thought of only as a global determination. Suppose, for example, that a social fact, like the distribution of the individuals of a society into classes of ages, can be described by a model whose first level will be a purely statistical scheme of urns, like that in §4, and whose second level is an 'informational' schema of the political and economic organization of the society. It would be an epistemological retreat to interpret either of these levels as a simple reflection. The progress of scientific analysis consists precisely in the formulation of the complex interplay of the system. Moreover it must not be thought that the opposition of our two systems can be presented as static and definitive. The very 'informational' part of a model can be dissociated by the progress of knowledge, and this cleavage gives rise to a new stage of 'control' in relation to which the remaining structure plays the role of energetic substructure. The contrast here is dialectical and by no means classificatory.

Such a conception may offer us a better understanding and a better application of the Marxist thesis of superstructures. Leaving aside the genetic point of view which I am discarding as being too far from our present purposes, would it not be possible to conceive the relation of economics to other social phenomena as the relation between 'energetic' structures and 'informational' structures? This would make more precise the interacting relation between infrastructures and superstructures regardless of operational mode. To recapitulate and summarize the results of our analysis, we may say that 'informational causality' is characterized by two traits:

(1) Recourse to the postulate of random strategic behavior. In game-theoretic models, the behavior of players at each of the stages is posited neither as a result that is energetically determined, nor as a realization approaching determinate behavior. Whereas randomness is introduced in a homogeneous

model as an uncontrolled residue of the determination of variables, it is completely different here for it becomes the instrument of this control itself: it is an integral part of the *solution* of the game, the expression of a higher mode of determination (cf. § 5.26).

(2) Reference to a norm which assures the canonical behavior of the set of the phenomenon. Models of human facts in this sense become kinds of paradigms, canons for a type of action, whose validity is tested at once by the conformity of results to predictions, and by the actual sensibility of the variables designated as strategic.

Thus a problem of a new sort appears on the horizon of objective knowledge, which is really a *metaproblem*, although it remains on the level of science. It is the problem of the constitution of norms. Naturally it can be considered from the genetic point of view, and that is, for example, what Piaget has done in his fine work on the psychology of intelligence. Turning to the most pressing matter first, one could also try to determine explicitly and exactly the normative conditions of such a model or of such a family of models of human facts. It is precisely this attempt, considered within the scientific process, that we are now going to analyze under the name of axiomatization.

This is, then, an attempt at a technological classification of models which will lead us to understand how the activity of axiomatization is naturally introduced into the sciences of man. For it is in fact the heterogeneous models which, in a *sui generis* manner, call for an axiomatic constitution, and it was in connection with such models, still very poorly distinguished from homogeneous ones, that the axiomatic process originated historically in the marginalist theories of 'value'. Moreover the axiomatization characteristic of the sciences of man has a meaning and scope very different from those to which mathematics and the natural sciences have accustomed us. I shall try to elucidate this process.

6.9. Our plan is to outline the epistemological function of axiomatization in the sciences of man, and to emphasize its originality. It is commonly believed that every attempt at axiomatization in this domain is only an artificial and abusive transposition of the movement which has animated an essential sector of mathematics for almost a century. If it is true that this is the historical origin of the process of axiomatization, a superficial comparative analysis reveals that axiomatization assumes very different functions when it is applied to objects of the natural sciences, and such an analysis provides a glimpse of the meaning which it takes on in the human sciences.

Logicians properly distinguish between formalization and axiomatization. To axiomatize is to posit principles which constitute a coherent and sufficient basis for the deduction of all the propositions of a theory. To formalize is to reduce the language of a theory to primitive terms and to explicit rules of construction. It is worth noting that all axiomatization presupposes a certain degree of formalization in language: one cannot axiomatize a body of knowledge whose expressions are vague, loose, and overburdened with overdeterminations. It goes without saying that an as yet unaxiomatized mathematics satisfies this requirement by its nature. This is hardly ever the case for other disciplines, and the formalization of language plays an integral part in the axiomatization, indeed it constitutes its essential moment. To axiomatize a theory is to try to rid the concepts that have an empirical origin of their uncontrollable contents. This process does not involve reconstructing the quality and the experience considered as such, but requires the substitution, for the latent structuration in the acts of perception and thought, of an explicit structuration of abstract elements. Axiomatization does not exclude, but rather *neutralizes*, the concrete aspects of the objects, thus permitting the transition from a vulgar and immediate practice to a mediate and developed one. It would not be distinguished from the construction of models studied hitherto, if it did not introduce a new requirement, or, more exactly, if it did not emphasize a new aspect of structuration: axiomatization is a mode of rigorous *definition* of concepts, and even, as I hope to show, of determination of objective *categories*.

MEANINGS AND FUNCTIONS OF AXIOMATIZATION IN MATHEMATICS

6.10. The technique and the history of the axiomatization of mathematics has been studied in depth. I only want to underline its epistemological function, as a way of introducing the study of the extension of this method. As is well known axiomatization is applied to an already mature subject and takes as its matter an already well constructed theory. However, an examination of some of the principal branches of mathematics reveals that axiomatization responds to different needs, and that beneath the dominant motivations described in the previous paragraph, particular *leitmotifs* come into play, depending on the state and specific nature of each domain. Thus, in geometry, the 'naive' axiomatic elaboration of Euclid was constantly brought into question because of the Fifth Postulate, and resulted, in the nineteenth century, in an attempt at the dissociation of presuppositions.

Bolyai showed that an 'absolute geometry', without the postulate of parallels, is conceivable, coherent, and fruitful as the study of the properties of a domain of objectivity which is, so to speak, freer, but no less rigorously defined than the Euclidean object. The same is true for the non-Euclidean geometries. Axiomatization functions here as a *method for the variation of the object*, and of the construction of new objects.

Peano's and Hilbert's theme of an arithmetical axiomatization is different. In the first place, it concerns bringing out the *structural characteristics* of a set of mathematical entities. Clearly the same dialectic of dissociation manifests itself here, but it is apparently governed by the plan to look for the fundamental properties of operations defined on a set of elements, properties which are ultimately presented as a rigorous and possibly unique definition of the system. The axiomatizations inspired by Klein's idea share this same spirit. According to this idea, a geometry is determined by a group of transformations leaving certain 'entities' invariant. Pushing the abstraction further, modern algebra defines systems of any operations whatsoever. It is the development of this theme which permits one to make explicit a fundamental idea of the axiomatic: that of the closure of the system that it defines. We will come back to this idea, for its importance is decisive.

Let me close this brief review by noting a third motif of axiomatization, which concerns set theory in particular. This theory is not so much a branch of the mathematical tree as it is the very ground in which it is rooted. It is thus not surprising then that, since its inception, it has given rise to problems of axiomatization, since, by its very nature, set theory makes explicit the principles of an activity which presupposes it. Axiomatic formulation is no longer only a secondary elaboration, it is the method of discovery, the method *par excellence*. The dominant theme is that of *coherence*, since intuition, having been reduced to the most analyzed of objects, takes its first tottering steps here. Cantor and his successors tried to articulate a *body* of rules which constitute, so to speak, a *charter of* primary data [evidence] capable of founding a mathematics.

6.11. Without wishing to examine axiomatization in mathematics for itself, it will be sufficient for us to outline its complex meaning, and to comment on what are, for our purpose, its two characteristic traits.

In the first place, the idea of closure. A system of axioms must define a domain which is, in a certain sense, self-sufficient. But this intuitive and vague notion can be clarified by a very nuanced analysis. We know that in his *Ideas*, Husserl used such an ideal to define what he called a nomological theory, as opposed to empirical theories. In a theory of this kind, every correctly formulated proposition must be demonstrable or refutable on the basis of the axioms, so that "*the concepts 'true' and 'formal implication of the axioms' are equivalent*" (Husserl 1952, p. 205). Now, since Gödel, this characterization of closure has been applied only to exceptionally simple constructions. Mathematics, in the current sense of the term, does not constitute a 'nomological' system. Or, at least, if there were closure, there would also necessarily be contradiction. Thus one cannot subscribe without paradox to the Husserlian thesis that the characteristic of closure is the very hallmark of mathematics: "... every deductive discipline which rests on such a system is a *definite discipline*, or one that is *mathematical in the pregnant sense of the term*" (Husserl 1952, p. 205). It will not do to claim, as Trân Duc Thao and Suzanne Bachelard do, that this condition, although an unrealizable ideal, "nevertheless retains its full value" (Trân Duc Thao 1983; S. Bachelard 1957, p. 112). The axiomatization of mathematics demonstrates that mathematics diverges from this 'ideal' in an *essential* way. On the contrary, it must be admitted that the notion of closure is dialecticized in the transition from the simplest systems – the theory of propositions and the first order predicates – to more complex theories. No doubt, the vulgar idea of closure remains paradigmatic as a requirement for domination made by formal thought over a domain of objectivity, but it becomes effective only through a redefinition each time in ways which are adequate for the different types of structures.[6]

A brief reflection on the diverse aspects which it can take on will serve to improve our understanding of further applications of axiomatization. The majority of logicians combine, without distinguishing, two perspectives on the closure of a system, one semantic, the other syntactic. From the semantic point of view, the closure of a theory corresponds to a certain degree of univocity of 'concrete' interpretations authorized by the elements of its structure. To say, for example, as Suzanne Bachelard does, that a system is complete (in the first of the senses she enumerates) for a determinate domain, "if it permits the deduction of all the valid formulas in the domain" (S. Bachelard 1957, p. 120), is to assume an interpretation of the system in which one distinguishes unconditionally realizable concatenations[7] of objects corresponding to those statements of the system which are called 'valid'. Thus, closure signifies that the extrinsic, semantic feature of validity involves the intrinsic syntactic feature of demonstrability. Closure is thus relative to a certain interpretation. It is by dialecticizing this notion of interpretation, or 'model'[8] that logicians succeed in defining and establishing the closure of the first-order predicate calculus, and in showing the impossibility of a closure so understood for systems of higher order.

Again, in this way logicians introduce the notion of categoricity, or univalence, a semantic property of a system of which all the interpretations are necessarily isomorphic, and consequently, not essentially distinct.

But if one wishes to restrict oneself to a syntactic perspective, without leaving that same universe of the calculus, one can define closure in a different way. This is the case for the Husserlian requirement and also for a weaker requirement: the addition to the axioms of a proposition not demonstrable (nor refutable) by them renders the system contradictory. These properties concern the set of propositions of an axiomatic theory: in the Husserlian perspective they depend on an *apophantic*, while earlier properties depended on a formal *ontology*. That it may be possible to express the same requirement in both languages is a rather risky thesis apparently implied by Husserl's philosophy of logic. The claim seems, however, to be implicitly admitted by a logician such as Church, in his *Introduction to Mathematical Logic* (1956) where he brings out the semantic 'motivation' of the concepts of consistency and completeness, and then speaks of modifying these originally semantic notions, in such a way that they take on a syntactic character (Church 1956, pp. 108 and 109). Nevertheless, the adequacy of this procedure is, I believe, nowhere demonstrated, and in any case the distinction between the two perspectives is essential for an understanding of the multiplicity of the possible definitions of closure.

6.12. If axiomatization aims in a certain sense at constituting totally dominated and closed systems of thought, this is by no means, as we have just seen, in a static sense. The second point to emphasize is precisely this dynamic of the process of axiomatization. One might in fact believe that axiomatic reduction only achieves in its form what is an already created science, and thus it represents a sterile phase of understanding. In fact the transition from the implicit to the explicit cannot at all consist in a simple formulation. The notions present as operators before their formalization can only really attain this conceptual level through axiomatization. This reduction thus dissolves the illusion of a separated subsistence which leads one to view these objects of thought as things in themselves incomprehensibly reconciled among themselves. This reduction reveals the object of science to us as a network of which only the nodes appear. Mathematical understanding is based on it, but without one having to oppose the act of providing a foundation to the act of discovery. If it happens that at first glance a theory so founded is a theory completed, embalmed like a mummy for the museum of science, it also happens that from this synthetic and structured vision arises the need for a more elaborate analysis, for a variation, a *rapprochement* with other domains. It is a commonplace to forget this animating feature [of axiomatization] in favor of the academic aspect, a commonplace already out-of-date. Moreover, if the tendency to axiomatize is at work in other fields it is certainly not because of its power to render a subject sterile.

AXIOMATIZATION IN THE NATURAL SCIENCES

6.13. Since Newton, every great period of renewal in physics has brought its contribution into an axiomatic form. For, in this domain the moment of axiomatization is constitutive. Of course, it is not chronologically first, but it corresponds to a decisive stage in the acquisition and recognition of the scientific object. In the intermediate stages, scientific practice in a confused way takes account of characteristics and uncoordinated points of view. The axiomatic formulation enunciates their presuppositions and if possible constitutes them into a system. It determines the physical object as a *possibility of models*, it defines the variables and the functions in terms of those which one can choose to characterize the physical object, it indicates how these notions rely on experimental processes. In sum, it furnishes a categorial framework for deductions and for experiments. In this way the three axioms of Newton's *Principia* present the object of physics as a system of 'masses', whose motions are to be described and which are supposed to be gauged in relation to an absolute spatio-temporal referent. These masses, by their presence, are mutually modifying each other's accelerations. It is apparent that this axiomatic account by no means aims at an ideal of closure comparable to that of mathematics. It constitutes a framework which is wide open to experimental determinations; with respect to its very interpretation it often remains equivocal, as the later progress of knowledge shows. It is in this way that the Newtonian category of physical object, which is spontaneously thought of as a system of things exercising mutual actions at a distance, came to suggest in its application a very different interpretation that developed into the Einsteinian axiomatic system. In this account, the physical object is thought of as a 'field' endowed at each of its points with local properties that depend only on the coordinates of space and time, and in which finally 'things' themselves appear only as mathematical singularities of a reference frame. But by associating with each material mass a 'potential' of attraction, that is, a local property attached to the space which surrounds it and which it occupies, Newton had already undertaken such an interpretation.

It would thus be inexact to see in the axiomatization of the physical object the final and restrictive elaboration of a completed science. No doubt, this is always the privilege of mature science, for it assumes a sufficiently lucid analysis of phenomenal data, and a certain mastery of experimental techniques; but the unification that it proposes for the properties of the object is only a sketch, at the same time retrospective and prospective, and it would be a mistake to insist that its role is only retrospective.

6.14. Let me briefly clarify this role: Axiomatization contributes, on the one hand, to the destruction of the prejudices of evidence, and on the other, to a sharpening of the relations between symbolism and experience. The destruction of the prejudices of evidence is assuredly one of the decisive aspects of conceptual thought. There is not one concept of physics that does not somehow assume the abandonment of such a prejudice; and, from this point of view, axiomatization may be defined as the substitution of a *simple idea* for a *common-sense idea*. This is a statement which is only apparently paradoxical. Common-sense notions, bound to an anthropomorphic and substantialistic interpretation of perception, can be simplistic; they rarely are simple, because they involve unexpressed elements, unformulated 'judgments', that constitute a stock of evidence at the perceptive and mythical level of daily life. The constitution of scientific objects always requires a revision of this evidence, and axiomatization represents the conscious and rationally developed phase of this process. The 'simple' ideas play a determinate role there, as Destouches noted (Destouches 1953, p. 28) but he added immediately that the notion of simplicity is "in large part subjective". This subjectivity is nevertheless only apparent. Simplicity has an intrinsic epistemological meaning, which is clarified precisely by the process of axiomatization. An idea is simple, I suggest, when it is introduced in a structural context, as opposed to an idea which is isolated. The idea of 'force' in Newton is simple, to the extent that it is connected with the ideas of mass and acceleration; similarly, the idea of entropy is simple, defined as total differential, because it is inserted in a mathematical model which coordinates temperature, energy, pressure, and volume. But still simpler is the idea of entropy as rethought by Boltzmann where it is an index of the probability of a state, for here its integration into the theory is more complete and more refined. We might call this conception of simplicity 'Pascalian' as opposed to the 'Cartesian' conception of '*natures simples*', whose characteristic is precisely that of being isolated. The axiomatic analysis of physics denounces the false privilege of '*natures simples*', and installs the reign of concerted [and systematized] primary data.

The other role of axiomatization consists in making precise the relations between scientific symbols and experimental techniques. The axiomatic formulation requires rigorous conceptual development, which substitutes simple ideas for rich and vague ones. But on the other hand this refinement elicits and makes possible an exact operational definition of notions, at the risk of making them appear too obviously as pieces of an imaginary construction. All physical axiomatization thus involves, in addition to the symbolic apparatus that constitutes it, semantic rules that connect it to experimental results. Thus, wave mechanics, when introducing ψ-waves as abstract elements of a mathematical model, matched them to the two principles of *spectral decomposition* and *interference* which express a relation between the intensity of the wave and its harmonics with the probability of the localization of, and the energy level of a particle. The probabilistic formulation itself naturally includes axioms that bring together mathematics and experience. It seems thus that the distance covered by axiomatization in the direction of abstraction is immediately compensated for by the possibility it provides of an increasingly accurate adjustment of the data of experiment.

AXIOMATIZATION IN THE SCIENCES OF MAN

6.15. The preceding analyses have shown us the plurality of the functions of axiomatization, in mathematics as well as in the natural sciences. It has enabled us to refute the still too widely held prejudice that axiomatic systems have a purely static and expository role. If it is true that an axiomatic system satisfies one rhetorical and esthetic requirement of scientific discourse, this is nevertheless only an accessory aspect, which must not obscure the dynamic and dialectic one that I have attempted to outline. Now, it is just this last aspect which, in the present state of science, explains and justifies the attempts of psychologists, sociologists, linguists, and economists. To the extent that, having moved away from the mathematical paradigm, one penetrates further into the experiential domain and one approaches the historical paradigm that dominates the scientific understanding of man, the instrumental and heuristic character of axiomatization becomes accentuated. Surrendering any pretension to a largely synthetic organization of the object, the axiomatic enterprise becomes essentially a means of *local research* which can only constitute its object piecemeal. While the *epistemological space* of the natural sciences approach the Euclidean type, that of the sciences of man seems to be in essence Riemannian. For the former, we can draw extended maps which immediately reveal its global structures. For the latter, only local explorations appear to be effective, leaving open the problem of relating two schemes of 'neighboring' regions. It would be quite imprudent to decide whether or not it will be possible to formulate a connecting law which would one day unify our knowledge. All that can be said is that today our knowledge progresses only in this groping way, which is perhaps essential for it. In these circumstances an axiomatic formulation cannot be presented as the end point of a synthetic process, a relatively stable stage in the evolution of an advanced science. Axiomatization is effective and justified from the outset of research, it is the instrument of discovery and testing. By its means, a concept that one wishes to test can be made precise and placed in a provisionally outlined context. The functions that we have noted in the domain of physics – destruction of pseudo-evidence and the [provision of] experimental articulation – here converge to rectify embryonic scientific thought, which is too easily blinded and confused by the brightness and glitter of experienced meanings. These attempts at axiomatization, however awkward and partial they may be, awaken thought from its repose in common sense. They offer themselves as explicit experiments of *eidetic variations*, carried out on initially shapeless notions, from which the minimal conditions of coherence and efficacy are extracted.

6.16. A brief example can show this remarkable aspect of the process better. We have already encountered several times the modern concept of quantity of information, in the form treated in the theory of communications and language (*cf.*, e.g., Sections 2.16 and 2.17). For a given symbol, quantity of information is defined as proportional to the symbol's probability of appearance and to the logarithm of this probability. This definition was specifically adapted to an envisaged phenomenon of communication and a certain amount of arbitrariness in the choice of the function adopted must be noted. But one can ask oneself whether a set of *minimal requirements*, which would characterize more radically the concept of information considered, does not correspond to this particular determination. Responding to this question involves formulating an axiomatic system that outlines only the mesh of a network, and leaves free a certain field of variation which will permit its adaption in neighboring domains of various specific constructions. This is just the sense of Schutzenberger's approach, based on a comparison of two apparently distinct concepts of information (Schutzenberger 1951).

The first is already familiar to us, the second is due to the statistician Fisher, who introduces it quite naturally in connection with the problems of the estimation of a parameter. This sort of estimation involves defining a measure of the information carried by a set of observations bearing on a random phenomenon. The quantity of information is connected to the accuracy of the estimation, and Fisher makes it a function of the *variance* of the estimated magnitudes.

A very simple axiomatic reduction illuminates the profound identity of the two points of view. The structure common to the two phenomena under consideration is the following: an observation is made on a random variable ξ susceptible of several different states A_i with respective probabilities a_i. This observation bears information in the sense that it permits one to decide whether or not the value of ξ observed belongs to the set X such that probability $(\xi \in X) = x$. A measure of this information ought to be adequate for this still vague, intuitive notion. To axiomatize is to make precise the formal requirements asked of the concept. Schutzenberger proposes that:

(1) The *functional* $H_{(x)}$, which measures the quantity of information carried by the observation of ξ, mentioned above, be uniformly *continuous*.

(2) The function be *symmetric*, that is, it amounts to stating that ξ belongs to X or that it belongs to its complement (corresponding to the probability $1 - x$); this feature expresses the fundamental dichotomous character of all information.

(3) The function be commutative, in the sense that it makes no difference to determine first whether ξ is in X or not, and then whether it is in Y or not, or to reverse the order of the process. This guarantees the additivity of successive information.

Such a system of requirements is coherent: A mathematician can easily draw from it the form of the functions that can satisfy it [9] and that depend on an arbitrary linear operator. The two preceding definitions thus appear as different specifications of the same concept, for two different choices of this operator. Of course, other choices remain possible, with the consequence that the axiomatization has outlined an invariant and rendered intelligible its apparently arbitrary eidetic variations. Thus it is clear that the formulation of an axiomatic system is not so much a process of rhetoric, as an instrument for the comprehension and generation of concepts.

6.17. One of the notions which are most fit to illustrate this function of axiomatization is no doubt that of norm of behavior. In Chapter IV, we encountered the idea of 'decision-making process' as the fundamental concept of operations research. In the present chapter (§§6 and 8), I have just emphasized the importance of 'rational behavior' for the human sciences. No matter what point of view or technique adopted by the analysis, it always involves the introduction into science of the objective idea of a norm of behavior. Neither the word 'decision' nor the word 'rational', which here simply symbolize two points of view, should make us believe in a return to a naive psychology of consciousness. It is not to Reason or to the Will that the psychologist, economist, or sociologist appeals. They claim they are describing and explaining phenomena by means of objectively constituted structures; normativity remains immanent to the model. In these circumstances the axiomatic analysis of norms appears to be the only way that is imposed. It alone permits us to radically eliminate the obscure implications of common sense, which cause the themes of scientific objectivation to intersect constantly with the interpretative themes of a philosophy.

THE EVALUATIVE STRUCTURE OF RANDOM SITUATIONS

6.18. Let us consider first of all the scheme of behavior of a subject when the personality of the 'subject' is reduced to that of a 'center of decision'. These decisions are taken in the context of environmental events, and they conform to a certain norm. Two fundamental hypotheses can be outlined on the basis of work by economists and psychologists: on the one hand, an *evaluation* of situations – rather than objects – must be assumed to be possible; on the other hand, the succession of events and their prediction by the subject should be thought *uncertain*: the behavior is carried out in the presence of risk.

Axiomatic theory can only aim at bringing out a certain structural coherence of this behavior. It cannot make a direct claim to a strict empirical description of effective actions, nor to the constitution of a framework of choices. But it furnishes the indispensable frame of reference for every experience and for all attempts at planning behavior.

The problem thus comes down to the construction of a decision-making process between different possible situations. Let us examine a 'situation'. It is presented to the subject as a group of different anticipated satisfactions, each one being assigned a different coefficient of probability. These probabilities obviously depend at one and the same time on the present state of the environment, on the laws of its evolution, and on what the subject will do. The decision consists finally in choosing among the situations enumerated. Let us call 'perspectives' the situations constituted by the exhaustive groupings of 'satisfactions' matched with the expectations of which they are the object. A decision is generally possible only if the set of perspectives is ordered; the preliminary question is therefore, what conditions can guarantee the construction of this order?

The most common tendency in all branches of the human sciences is to postulate the possibility of setting up a correspondence between each perspective and a *number*, by means of a simple function. This number furnishes an index of the order of preference of situations, and eventually even a measurement in a sense which must be made precise in each case.

From the beginnings of the calculus of probability, this problem has been posed, though indeed in a less general sense. Two classical solutions were immediately formulated. One is Pascal's function: mathematical expectation, which, assuming the *values* of the satisfactions to be objectively defined, is calculated by adding the products of these values and their respective probabilities. These probabilities themselves, whatever their meaning and their manner of estimation, are supposed to measure the expectations of the subject. The other solution is the function of Daniel Bernoulli, which Laplace called 'moral wealth'; it is calculated like the former function, but replaces the objective measures of the values by their 'subjective' ones, which depend on the initial wealth of the subject. Bernoulli proposed to determine these measures as proportional to the logarithm of the objective values (Bernoulli 1730–31).[10] Not to be confused with the objective value, subjective value increases less quickly and can be represented in Cartesian coordinates by an inflected logarithmic curve. One can speak of a *curvature* of satisfaction in relation to objective values. Of course, the axiomatization of perspectives would leave indeterminate the choice of the law of curvature which best suits the results of observation.

6.19. But if one wishes to give a precise sense to a calculus of these perspectives, as economists in particular want to do, it is still necessary that our axiomatic system make explicit different properties of the numbers defined by the function. This is the goal pursued by von Neumann and Morgenstern in their *Theory of Games* (1944). They begin with a pair u, v of incompatible satisfactions, matched with expectations α and $1 - \alpha$, and consider only the dichotomous perspectives of the form $\alpha u + (1 - \alpha) \cdot v$. The + sign used here designates the original operation of combining two elementary perspectives, and the axioms give it the formal properties of algebraic addition. The two most remarkable traits of this axiomatization are those which establish a certain continuity of perspectives and the substitutability of an alternative for an elementary perspective within another alternative.

The first feature is formulated as follows: $u < v < w$ implies the existence of a probability α, such that: $u < \alpha \cdot u + (1 - \alpha) \cdot w < v$ (which results from 3.B. *a* and 3.B. *d* in von Neumann-Morgenstern (1944, p. 26); one could equally well have the dual formula, employing '>'). This means that between two satisfactions u and v, a perspective of intermediate value can always be interpolated, by introducing any satisfaction w preferred to v, and a suitably chosen expectation α.

The second feature involves the possibility of replacing in the perspective

$$\alpha u + (1 - \alpha) \cdot v$$

the satisfaction u, for example, by an alternative of the form

$$\beta u + (1 - \beta) \cdot v$$

such that the result is an alternative in which the coefficients α and β obey the rules of arithmetic:

$$\alpha[\beta u + (1 - \beta)v] + (1 - \alpha)v = \alpha\beta u + (1 - \alpha\beta)v.$$

Thus, one can reduce a complex perspective to an alternative, a ticket in a lottery where the prizes are themselves tickets to a lottery with a simple ticket . . .

These requirements which condition the reduction of perspectives to numerical scales are not, however, free from difficulties. They introduce, as representing expectations, coefficients which have the mathematical properties of probabilities. But it may be asked whether the evaluation of perspectives would not be more adequately realized by somehow considering expectation and satisfaction as connected; so that, for example, in the neighborhood of certainty – and of impossibility – the estimations of satisfactions are over-estimated or underestimated according to a determinate law. In a region far from these critical zones, everything would happen as if expectations and satisfactions were independent. Thus it may be necessary to conceive of a sort of relativistic schema introducing a 'curvature' of the satisfaction-expectation *variety* in a sense analogous to that of Einsteinian space-time. Such a refinement of axiomatization has not, I believe, ever been attempted. Nevertheless, it is the type of problem which an attempt to axiomatize the structural concept of evaluation of a probabilistic situation poses. If we are to characterize a decision-making process completely, we must still define a principle of choice.

THE DEFINITION OF A NORM OF DECISION

6.20. This idea is presented with complete clarity only by the great reformers of economics at the end of the nineteenth century, under the still very rough form of a rule of choice among different goods, or among uses of a good available in limited quantities. The schema was originally static, in the sense that the goods to be chosen were given, and the satisfactions and the costs corresponding to them were immediately and exactly predictable. The norm very naturally adopted thus the maximization of the total net satisfaction, and the technical procedure deriving from it under these conditions is the equalization of marginal satisfactions.[11]

But the schema of decision is modified in a more realistic sense by its reduction to a *game* structure. If one in fact considers that the result of a choice depends not only on the choice, but also on a random response of the environment, the attitude of a subject can be assimilated to that of a game player. The simplest situation is describable then by means of a matrix or function of pay-offs, furnishing the value of satisfactions which result from the combination of each of the possible choices of the subject and each of the types of responses of the environment. The latter intervenes in the schema as a 'player' to the extent that the types of responses which constitute its 'tactics' are not predictable.

Subject \ Nature	T_1	T_2	...	T_j
T'_1	s_{11}	s_{12}	...	s_{1j}
T'_2	s_{21}	s_{22}	...	s_{2j}
T'_3	s_{31}	s_{32}	...	s_{3j}

T_k : tactics of nature;

T'_k : tactics of the subject;

s_{ij} : subject's pay-off for a T_i and a T_j;

p_n : probability of the tactic T_n of nature.

In this perspective, a more elaborate axiomatic analysis leads to two sorts of models. In the first, the environment's tactics, although unpredictable, intervene according to known probabilities. In this case we could speak of *Bayesian models*, since it is one analogous hypothesis which establishes the so-called theorem of the probability of causes.[12] In the models of the second kind, no *a priori* probability of the 'adversary's' different tactics is postulated; one simply knows how to count them, and one knows the matrix of the payoffs. In the Bayesian models, a norm for decision-making, derived from the notion of mathematical expectation, or the average value of the payoff, is, quite naturally, imposed. The subject chooses from among his tactics the one which will maximize the mathematical expectation of his satisfaction, that is, the linear formula: $\Sigma p_j s_{ij}$, defined for each T_i'. If the nature of the phenomenon is such that the decision can be repeated a sufficient number of times in the same conditions, such a norm effectively insures the maximization of the average satisfaction, once we admit the principle of the adequation of schemas of probability to experience. But this norm is obviously no longer applicable to models of the second sort, in which the p_i's are not postulated. In this case the simplest solution for the subject appears to consist in choosing that tactic which is *capable of providing* the greatest satisfaction. This tactic obviously 'finesses' a possible dominating tactic of nature. If this dominating tactic is used, the resulting satisfaction can be very small or even minimal.

6.21. In any case, another norm has been defined, which constitutes the fundamental theme of the theory of games. Decisions are to be taken in a way that assures the highest satisfaction compatible with the tactic of nature most unfavorable for the subject:

$$s' = \max\,(T_i')\,\min\,(T_j)\,s_{ij}.$$

But the existence of such a tactic, called *maximin*, depends on the nature of the matrix of pay-offs.

In fact, the intuitive and vague notion of 'the most unfavorable tactic of nature' has no precise meaning, since no hypothesis can be formed on th way in which the environment determines its own tactic. In order to give it a meaning, the hypothesis of the 'evil spirit' must be advanced. Nature is supposed to choose its tactic in order to minimize the satisfaction of a subject searching for the most favorable tactic, or:

$$s = \min\,(T_j)\,\max\,(T_i')\,s_{ij}.$$

If the two values s and s' of s_{ij} thus determined coincide, the subject's norm of action obviously has a univocal meaning and one can easily show that every other tactic can involve a lower level of satisfaction.

If the values do not coincide, no pure tactic has an optimal character, and the norm of choice is indeterminate. This is where von Neumann's theorem comes in, establishing the unconditional existence of a solution in a new sense. The norm of action is no longer a pure tactic; the subject must have recourse to a mixed tactic, or strategy according to which his choice must be made between different pure tactics, each of which is assigned a coefficient of probability which the theory enables him to calculate. Thus randomness is reintroduced, but no longer as an objective feature of the environment; here it becomes the constructed feature of a rational behavior, an instrument of a normative technique. I shall not return to this dialectic of probability in the human sciences. (For this, cf. Section 5.16ff.) My present purpose is to illuminate, by the example of decision making, the fruitfulness of axiomatic analysis.

6.22. Clearly, the complexity and abstractness of such an apparatus are nevertheless striking. Undoubtedly in this form the axiomatic construction of a model of decision making is still very far from effecting a union with the mass of actually observable phenomena. But it must not be forgotten that such research is meaningful only if it retains its *local* character. The aim is not a general theory of decision but first and foremost a partial structure of certain typical phenomena. This must be seen not as a very general categorial determination of the human fact but rather as the attempt to provide a conceptual technique; and from this attempt there may develop, sooner or later, the general idea capable of establishing finally the status of the human sciences.

On the other hand, it is on the basis of an axiomatic analysis oriented towards the construction of a complex and rigid model that less ambitious though immediately adequate models can be formulated. In a chapter of the work of Simon previously cited (1955a in *Models of Man*, 1957), this desire to abandon hypotheses that are too precise is clearly expressed. The author then designates several points which could be made more flexible in a weaker model: reduction of the scale of satisfactions to two or three discrete values, abandonment of the postulate of a total ordering of satisfactions, rejection of the hypothesis of complete information on the part of the subject about the set of pay-off matrices. The norm, which has always been presented as the search for a maximum, could be itself weakened so that it consists in search for a satisfaction *not lower* than a given level of aspiration, which could moreover be varied as a function of actual information. This is the case for La Fontaine's Heron [*Fables*, Livre VII, Fables IV and V], whose level of aspiration decreases as his knowledge increases, through successive failures in his search for an optimum.

But this weakening or softening of a structure can only be meaningful, and rightly so, through axiomatization. It is axiomatization which reveals the interdependence of hypotheses, and their strategic value in a model. Even when it cannot be elaborated to the explicit and rigorous form required by mathematics, it calls forth and assumes this eidetic variation of models that is one of the essential conditions of the construction of effective and coherent concepts.

CONCLUSIONS: CONSCIOUSNESS AND CONCEPT

6.23. The movement towards axiomatization in the sciences of man thus should not be confused with a tendency to rhetorical formulation. Nor, moreover, can it simply be reduced to a search for constructive rigor, a search which is necessarily belated, and which flourishes only in the most advanced and abstract sciences. No doubt, in the human sciences axiomatic formulation *also* plays the roles which we have tried to discern in the other types of knowledge; it has the function of closure relative to an objective domain, a 'doxolytic' function for the neutralization of prejudices, and the function of clarifying the semantic problems posed by a symbolism. But in the human sciences axiomatization has a special value which makes it the instrument not of an advanced science but of a developing science. It would be quite wrong in this domain to distrust attempts at axiomatization of a discipline moving tentatively, for here axiomatization is not vain ambition, but a necessary step. If the axiomatizations proposed by the psychologist, the economist, are of any use for the progress of science, it is not because they appear to ape the constructions of mathematics. It is because they offer to rational thought the sole means of escaping from the attractions of data derived from experience. In the domain of man, the immediate meanings which constitute the natural mode of the presentation of phenomena risk the total concealment of positive structures, the only determinations possible for an object of science. This magic is obviously not at work to the same degree in the other fields; naive mathematics reaches a certain spontaneous rigor easily enough, even when it remains charged with a symbolist metaphysics which never radically obliterates the structural determination of the concept. Thought about physical phenomena, although more threatened, also separates itself, prior to every explicit axiomatization, from the confused implications of concrete experience and ideology. In both cases the axiomatic reduction, if it is the origin of new progress, is first of all an achievement, a point of arrival. "You would not search for me" says axiomatic science to naive science, "if you had not already found me." It is generally no longer the same in the sciences of man. The attempt at axiomatization is here a preliminary groping about, the necessary preparation of a field of operation, by means of a drastic and certainly aggressive asepsis of common notions. Moreover, we must on the other hand see in these constructions, in the present state of affairs, only the eminently provisional scaffolding for the construction of concepts.

6.24. Such is the particular epistemological status of an axiomatic system in the sciences of man. This original dialectic, moreover, only brings to light the very nature of the most general conceptual thought. For the scientific concept, whatever its domain of objectivity, can be defined in the final analysis only through an axiomatic movement. Every notion effectively utilized by science is a notion *on the path* of axiomatization, or an axiomatized concept. Thus, for the most perfect examples of triumphant conceptual thought we must turn to mathematics, and in order to find examples of militant and long-suffering conceptual thought we must turn to the sciences of man. In the mathematical domain notions are spontaneously treated – if not thought – as structural complexes which axiomatization reveals in a sort of apotheosis. This is the universe of 'Grace'. The domain of human facts, scientifically speaking, is the universe of 'sin'; there must be a conscious will to axiomatize in order to lay bare the concept. But in both cases, this fact is disengaged from the experienced, confused, subjectively centered notion only by means of the axiomatic reduction of a structure.

My generation, which has been more or less acquainted with the work of Jean Cavaillès, gladly quotes and comments on his judgment, a program for a *philosophy of the concept*, as opposed to *philosophies of consciousness*. In fact this makes a great deal of sense for an interpretation of knowledge. For my part, I understand it as first a rejection of idealism, but also and above all as a transformation of the paradigm of knowledge. I think that 'consciousness' is, epistemologically speaking, an operatively isolated act, based on itself, having an *essence* for its correlate and clarity for its quality. The concept is the systematization of operational acts, having for a correlate an explicit *structure*, and for quality, coherence. Consciousness designates a mode of experience centered around the *Ego*, and the concept designates equally a mode of experience, but one which is decentred, organized, and open to a possible hierarchy of evident clarities.

That consciousness is an irreducible mode of experience, and uniquely that of human things, cannot be legitimately contested by any philosophy of science. This consciousness certainly is part of *praxis*, understood as the most complete and concrete experience, but also one which is constantly modified, recast and threatened. A 'philosophy of consciousness' is wrong only to the extent to which it tends to erect consciousness into an integral and definitive experience. For at the heart of *praxis*, consciousness represents only the illusion of stability, which responds to our need for the absolute. An *aliquid inconcussum* cannot be rationally discovered outside of consciousness, except through the intermediary of consciousness itself. But then it is necessary to shut oneself out of the world of practice, and ultimately to abandon rational thought. A philosophy which makes an effort to remain rational cannot remain a philosophy of consciousness, in whatever sense it may take. We still have before us the counter-example of neopositivism (with Wittgenstein's itinerary among others), and that of phenomenology.

6.25. In aiming at a philosophy of the concept, I want to remain a rationalist – or rather, as Gaston Bachelard says, I want to try to become one. It is not necessary to believe, however, that such a perspective, such a defiance of 'consciousness' and of the 'essences' it aims at, involves an irrevocable abandonment of the transcendental point of view. On the contrary, the study of scientific thought confirms the thesis that every concept remains philosophically incomprehensible unless it is grasped in its transcendental character. In the activity of conceptual thought, at least three inseparable instances must be distinguished – *instances* rather than moments, for each of them can be present without a predetermined dialectical order commanding its appearance. There is the idealist instance according to which the *form* itself is taken for the object and confused with the object; there is the realist instance where the notion becomes a tool and where science tends towards a technique; there is the transcendental instance where the object-structure is related to its conditions of validity, which are neither the forms or norms of a subjectivity but the explicit rules of a certain provisional arrangement of experience. One will doubtless say that the meaning of the word is diverted, and that the interventionist *ego*, being in the world, would not leave any really transcendental traces in its work. I continue to employ the word, however, because it seems to me to preserve what remains alive and true in the Kantian analysis. Namely, the subject gives himself rules for the game of understanding, and he constitutes the object. But it has seemed to me that this constitution was a *labor*, and not the definitive and free gift of an inexplicable nature outside of nature.

Thus, what seems to me to remain of the really transcendental in scientific activity as I have described it, is the constitutive – although provisional – and aprioristic position – although issuing from a long labor – of the categorial sketches of the object: that is, its axiomatic aspect. But this axiomatic aspect, I have attempted to show, is the other side of a pragmatic aspect, and the human freedom to which it testifies loses all its meaning if one wants to isolate it from this codetermination, through which man is bound to a world which makes him man, and the world to man which makes it a world. In this way, both the transcendental requirement of a thought, aiming to constitute itself as an *object* which is neither a simple impression, nor even a percept, – as well as the compelling conditions imposed by a world on its activity of appropriation – become obvious.

6.26. If science, and even the science of man, is really an activity involving concepts and the construction of concepts, then it faces the problem of the individuation of its objects. In a philosophy of consciousness the grasp of the individual is not a problem, for every act of consciousness, appropriately analyzed, always results in the perception of the individual. It is in taking such a philosophy into account that Aristotelianism, which is, however, a prelude to a philosophy of the concept, proposes this strange thesis: perception – *αἴσθησις* – as a faculty, is of "the such" – *τοῦ τοιοῦδε* – and not merely of a "this something" – *τοῦδέ τίνος* –; yet one must at any rate actually perceive, with respect to the act of perceiving – *τὸ αἰσθάνεσθαί* –, a "this something", and not a definite present place and time (*Anal. Post.*, 87b 29). That is to say that, if perception, isolated, as a component of knowledge, furnishes us with *qualitative abstracts*, perception as a *complete* act of understanding, as consciousness, has for its object an individual. This was in fact one of the further reasons which in Aristotle's eyes disqualified perception as science. Essences, according to the philosophies of consciousness, appear rightly as mediators between the individual and the concept. This is a mythical mediation, for it conjures away both the dialectical nature of the concept and consciousness by making them revolve around a fixed imaginary point.

However, in a philosophy of the concept, the perception of the individual is not free from problems, especially in the sciences of man. Hegel himself, who first explicitly gave philosophy the yearning for a theory of the concept, but who returned, in the *idea*, to a philosophy of consciousness, encountered the difficulty in his chapter on reason in the *Phenomenology* (Hegel 1977, pp. 139–262). Observing reason can grasp the individual only through signs, an expression, an externality. It can then only grasp it as something alienated, and finds itself led to the paradoxical thesis underlying physiognomy and phrenology: "Spirit ... is a *Thing*, ... the *being of Spirit is a bone*" (Hegel 1977, §343, p. 208).

It is observing reason which, as Hyppolite remarked "[isolates] exterior and interior and then [claims] that they correspond" (Hyppolite 1974, p. 268). It ignores the dialectic of the consciousness of the active self, which grasps individuality not in the alienated work, but in the operation itself. On the basis of this, however, theoretical activity is achieved and surpassed; the practical moment is in operation, in which the thing is no more than an immediacy having to be suppressed. According to the Hegelian expression, we see, "disclosed the *realm of ethical life*" (Hegel 1977, §349, p. 212).

Must we in the end renounce a *science* of the individual? The response of Hegelian idealism is affirmative, precisely because it is an idealism. Affirmative, at least for the meaning we have given the word 'science', for the Hegelian response might well be that in the final analysis knowledge of the individual is history, but a history in *ideas*, and not in concepts, this word being taken with its Hegelian meaning, and the restriction is naturally valid, *a fortiori*, in the sense that we have given it in this chapter.

We should ask ourselves about the possibilities now offered to a science of that 'active reason' which Hegel referred to the realm of ethics, and which he considered only in a philosophy of history. For this appears to be, for the sciences of man, the supreme critical test: to what extent, by what means, on what foundations, can these disciplines hope to arrive at an objectivization of the individual.

CHAPTER VII

THE UNDERSTANDING OF THE INDIVIDUAL

7.1.

Speech and work are outer expressions in which the individual no longer keeps and possesses himself within himself, but lets the inner get completely outside of him, leaving it to the mercy of something other than himself (Hegel 1977, § 312, p. 187).

So wrote Hegel in the chapter on observing Reason to which I made allusion earlier. Language and work are two forms of individual expression which he considered as "[expressing] the inner" too much and too little. Too much because they do not permit the existence of any opposition between them and the individual; too little because the interior, in its very expression, is altered, and turned into something else. Under these circumstances, one is ultimately reduced to abandoning the individual to the consciousness of an irreducible experience, transposable only through the marvels of the work of art.

But neither work nor language are immediate activities in relation to the subject; thus it cannot be said that they suppress an opposition between themselves and an interior, since they, on the contrary, represent the *determinate* modes of opposition between a subject and the world. It is the alteration and alienation that they assume, which gives meaning to this 'interior'. As Hegel said himself, " . . . action is simply the coming-to-be of Spirit as *consciousness*" (Hegel 1977, § 401, p. 240). It remains to be seen whether a science of this *active reason* is possible. From our non-idealist perspective, we see that such a reason is concrete objectivity and by no means appearance, that it is the very objectivity of human existence.

The persistent tendency of idealism is, at bottom, to turn the individual into an epiphenomenon: the springs of action would be elsewhere, at once short of and beyond the individual, such that in open or hidden forms the idealist theories of the human world are occasionalisms. Individual fate can scarcely but be referred to an esthetics or a theodicy. If one insists on introducing into one's purpose a semblance of objective knowledge, one must appeal to *chance*, but to a chance considered as a mysterious residue of theology and teleology, not to that concept of chance as an instrument of an effective description of acts, which we have encountered at every point in our attempt at analysis.

Now, there is the most dangerous confusion reigning in the domain of the sciences of man owing to the fact that the very people who profess this negative doctrine, hypocritically reproach science for being stumped by the problem of understanding the individual. I intend to denounce this cunning ruse of irrationalism.

7.2. The status of knowledge of the individual is certainly the major difficulty for an epistemology of the human sciences. But the problem cannot be solved by systematically denying its possibility, or by refusing all objective consistency to the individual. At first glance we find ourselves boxed in by a dilemma: Either there is knowledge of the individual, but it is not scientific – or there is a science of the human fact, but it cannot reach the individual. No brilliant success in psychology, or in sociology, has yet produced unquestionable proof of the speciousness of the alternative. But the fact is that the features of a new conception of science in this domain have scarcely been outlined and the ideal implicitly recognized by these criticisms remains inextricably bound to a state that is completed. If the understanding of man had in fact, to be speculative, then we would have to accept the terms of this dilemma, and consequently resign ourselves to seeing this understanding halted at the threshold of its true object, a science infirm and disappointing, general without an established universality, abstract without perfect rigor.

But if one admits the accuracy of the preceding analyses, if one recognizes in this science of man an increasingly lively activity, better and better formulated towards a status of applied knowledge, it will be seen that this science escapes the dilemma. A *speculative* science of the *individual* is impossible, it is true: this is the meaning of the Aristotelian aphorism, that there can only be a science of the general. But as soon as a science succeeds in governing a *practice*, and tends to be constituted as an integral *praxis* in its own domain, it addresses the individual. Clearly, it is in the world of human facts that this promotion of the understanding takes on its most significant characteristic, and encounters the most redoubtable obstacles. Because, in fact, discounting all anthropomorphism and all mythology, the very notion of the individual is constituted, on the plane of the natural sciences, only at levels of weak complexity, under almost anodyne forms. That chemists and physicists, whose science is now deeply integrated into practice, work on objects infinitely more organized and technically determined than the science of Lavoisier and Newton, would hardly shock anyone. In these new objects, in these 'effects', in these highly differentiated products of a rational applied science, we do not recognize the archetype of what is abruptly becoming, in the human domain, the individual considered as an object of science. Here is a gap, a rupture, no doubt, but it is true nonetheless that it is the very same process of integration into a practice that opens the way to an ultimately scientific conception of the individual. In order to characterize this conception one can speak of the '*clinical*' *pole* of the sciences of man. The word remains obscure, because it is charged with undertones inherent to a specific practice, medicine, which, to my mind, has still not broken sufficiently clear of its myths. I shall try to show how, by clarifying the notion of clinical, it can be stretched to the set of a practice which extends and penetrates all the sciences of man.

Hitherto, our attention has been directed essentially at the determination of structures as schemes of the object they refer to. We have now arrived at the point where the problem of a *linking up* of structures comes to the foreground, and where the function of thought must take on a dialectical sense in a decisive fashion. Based on the stages already reached by science, my reflections can naturally only sketch the perspectives of a development of its categories. I want to avoid prophesying, and wish only to outline the direction of the progress immanent in the actual state of things, for this is the end point and goal of a comparative epistemology. I shall approach my theme by examining the relations between the 'clinical' point of view, and structure, this time borrowing facts from different domains of psychology. We shall encounter the problem of history, but only, of course, through a determinate basis, that of the paradox of a *clinical undertaking without practice.*

Finally, by way of concluding this chapter and this study, I shall sketch a characterization of the human object through the categories of model and field, which will enable us to glimpse the direction of a scientific grasp of the individual.

THE CLINICAL SITUATION AND STRUCTURES IN PSYCHOANALYSIS

7.3. In the following pages we are going to approach the difficult problem posed for science by its relations to the individual. If it is true that the very movement of scientific thought consists in always opposing a process of structuring to formless data, and in pushing back the limits of the immediate, so to speak, it is nevertheless certain that science cannot indefinitely challenge direct contact with events, with the world. It must stop. Its movement is authentic only because it is, at any moment, capable of achieving in a practice something that can by no means be reduced to a simple experience in the traditional sense. In the domain which we shall consider, this contact with the concrete individual can be designated, as we said, by the word 'clinical', borrowed from physicians. But what is the clinical point of view? A psychoanalyst (H. Hartmann 1959) has noted quite rightly that this question is given no satisfactory answer in the philosophy of the sciences. We cannot hope to fill a similar gap; this would require a whole work of its own, oriented in a direction other than mine, one which would be a counterpart to this one. However, it is indispensable here to indicate what appears to be essential to the elucidation of this methodological category, though it is still imprecise and disfigured by its narrowly medical origins.

The clinical situation brings into immediate relation the patient and the therapist, the observer and the observed. By 'immediate relation', we must understand a relation not totally conceptualized, involving in an initially confused fashion the reactions of both, so that the situation which is established cannot be correctly described as a totally asymmetrical encounter between an active subject and a passive object, but rather as a couple of which both partners play alternating roles. It cannot be a question of a situation of speculative knowledge or even, originally, of applied *knowledge*. The clinical situation is spontaneously experienced in the magical and mythical mode of communication. The chief epistemological problem is to explain how this situation can be developed into a register of authentic knowledge, without degenerating into a crude technique of mechanical objectivation, or into a spell-binding practice. The history of medicine itself shows the diverse vicissitudes of the passage between these two reefs: between the curious 'mechanistic' excesses of Iatrochemistry, and the magical approach of Mesmerism; between the rigidity of the Pasteur's objectivation of disease, and the romanticizing tendencies of psycho-somatic theories. The very alternation of medical doctrines reflects the two possibilities in the attitude of the practitioner.

The situation of the sociologist, the psychologist, the economist, from the moment they assume the actual presence of a well-determined human reality, to a certain extent reproduces the same ambiguity. Another feature, however, seems necessarily to prompt the rejection of the analogy: in the physician's clinical attitude, the focus is certainly predominantly on the pathological. The patient is perceived and thought of as ill, and this perspective cannot be extended indefinitely without paradox. Does not the most fruitful epistemology of medical science result precisely from the dialectic of the 'case', from the complex and contradictory grasp of the singular and the symptomatic, from idiosyncrasy and syndrome? It is a dialectic which is theoretically very confused, but concretely effective, and its most lucid realization must be sought for, not in the Hippocratic texts, but in the Aristotelian theory of knowledge. This pathological element in the focus of the 'case' is not really essential, at least in the sense in which it is understood in the science of the physician. In order to establish the dialectic of the case, it is sufficient that the fact examined, that the second term of the clinical couple, be viewed as 'deviant' in relation to a schematic construction. And this is what appears to me to be the positive feature of the individual for science. The whole traditional theory of knowledge rejects precisely this notion of 'deviant' as escaping science. It was to this effect that Aristotle establishes his doctrine of the universal and the accidental. In the latter there is no science, but within all the epistemology of this philosopher there nevertheless holds sway a sort of yearning for the individual. *Callias*, for science, is only a shadow; but it is with him, however, that we deal, and one of the profoundest movements of Aristotelian thought leads to a supreme type of being, defined at once as universal and as a singular individual through the physical theology of the prime mover. But this reconciliation is too obviously ineffective. Early modern science, which sought to read nature as a book printed with mathematical symbols, had radically to renounce the perception of the individual; more Aristotelian in a sense than Aristotle himself, it was a science of the universal and of the universal alone.

The theory of probabilities in a certain way reintroduced the deviant, but only insofar as it represents a class, and not as an individual. At least it permitted the representation of the symbol of the individual in the schematizations of science, under the guise of a variable and as a blank space. But it is the transposition of the clinical situation to various disciplines dealing with man which makes the problem of an understanding of individual *contents* explicitly reappear. The object viewed as deviant is thus no longer only a system of possible variations, an empty shell. It is a question, as Lagache says, of

> envisaging behavior in its proper perspective, of setting up as accurately as possible the ways of being and acting of a concrete and complete human being at grips with a situation, (of) trying to establish its meaning, its structure and origin, (of) disclosing the conflicts that motivate it and the steps which tend to resolve them (Lagache 1969, p. 15).

Thus an epistemological malaise originates in the disparity between traditional norms of understanding, and the newly promoted type of understanding that one hopes to derive from the clinical situation. In order to grasp the development of this conflict, let us turn for a moment to the discipline which is its preferred place: psychoanalysis.

7.4. The demand for a psychoanalytic psychology holding the rank of science, if it goes back to Freud, is now an altogether pressing issue in certain analytic circles. The American L. S. Kubie (1959) even goes so far as to regret that psychoanalysis has been absorbed 'too soon' by therapeutic tasks, and has not been able to completely develop its methodology. For my part, I think that this therapeutic urgency, far from having hindered the theoretical progress of psychoanalysis, has constituted the only barrier which preserves it from divagations and myths. What can become a meta-psychology set free from therapeutic constraints is foreshadowed in the Heideggerian writing style of certain analysts. Rather than producing a solid theoretical and experimental construction, psychoanalysis without clinical practice would only engender a mass of verbose, obscure, bombastic and empty talk. Accordingly, it is by no means the therapeutic tasks which obstruct the constitution of a science in this domain. The difficulty derives, more radically from the very nature of the clinical situation on which psychoanalysis has had the merit to found itself. Because of limitations of space, and those of constitution of a science in this domain. The difficulty derives more radically psychology. I shall be content to outline, within the general perspective of the sciences of man, some of the characteristic problems posed therein.

7.5. In the first place, there is the problem of a balance between the conceptual and the experienced. It is well known that the psychoanalytic process considered as cure rests on the reproduction, as experienced by the patient, of conflict situations, and also on the discovery – also experienced – of their latent meaning. But it is quite obvious that analytic understanding is not, in Freud's view, identified with this direct grasp of phenomena, even were it the result of a long program of recollection and reconstruction. In *Psychoanalysis and Medicine*, Freud recalled, *à propos*, the anecdote of the applicant for a position of children's nurse: "Do you understand children?", she was asked. "Certainly", she responded, "I was once one myself." The consciousness of what has been experienced, brought to light by analysis, is one thing, and the scientific analytic understanding of the individual is another. One cannot say that psychoanalysis has succeeded in offering a clear and incontestable conceptual apparatus that can serve as a framework for controlled understanding. From this point of view, the vigorous and malicious criticism of Ernest Nagel (1959) remains pertinent. Psychoanalytic theories, he writes, are formulated in such a way that they cannot be refuted by facts. The criterion of coherence of an interpretation cannot suffice, for it is always possible, by some ingenious means, to find *several* coherent interpretations. As for the control obtained through the verification of the consequences of childhood traumas, it concerns traumas *currently recognized by the subject*, and not the traumas actually observed. But the current inadequacy of conceptualization in psychoanalysis should not rush us to a condemnation without appeal. The fruitfulness of its clinical point of departure derives not from the fact that it provides a solution, but rather from the fact that it poses a problem in a radical way, which is decidedly that of the transposition in objective and controlled understanding of the active grasp of a situation. It might perhaps be believed that what is involved is simply a particular case of the inductive approach. This is not so, for the fundamental theme of induction is the elimination of the individual – as deviant – while a clinical understanding aims essentially at the individual as such. What several psychoanalysts, reflecting on their method, seem to be pointing towards, is a precise conceptual determination of the nature of the analytic couple that constitutes the original category, the basis of this new type of knowledge. L. S. Kubie (1959), having shown the limits of the method (in the light of the traditional ideal of scientific understanding), tries, for example, to provide a functional description of the analysis-analyzed relation. As a result of the analytic incognito, the analyst remains for the patient an indeterminate human being. Projecting on him his own feelings and dreams, the patient can come to an awareness of the distortions of reality which these projections involve. The role of the analyst would thus be to permit a sort of reduction of the patient's psyche which turns it momentarily into an experience suspended outside of reality. On the other hand, to the extent that he is active, the analyst reattaches the patient to reality and obscures for him the role of feelings and unconscious needs. All things considered, the author hopes that the application of the new conceptual techniques of cybernetics, by permitting the construction of rational models, "will make much clearer what is today confused and obscure in psychoanalytic theory."

It is quite remarkable that, from a very different perspective, a French psychoanalyst (Bénassy 1957)[1] appeals to the notion of model in order to make precise the relation between psychoanalysis and psychology. It is in the convergence of the 'intuitive model' of psychoanalysis 'rendered explicit' and of a neurophysiological model that he sees the progress of psychology. This is certainly too narrow a conception, but it puts the accent on the need for a conceptualization of the experienced by way of structural analogies. I shall come back to the meaning and the particularities of such a treatment of the individual. For the moment, I shall conclude by emphasizing that the relation of the conceptual to the experienced in the clinical perspective apparently needs to be made more precise by the elaboration, in the form of a model, of the situation of the analytic couple: thus the absolute and the immediate of the individual can come to be dialecticized.

7.6. A second important aspect of this problematic is that of the function of language. The importance of speech in analytic therapy was emphasized even before Freud's decisive intervention. It was *Anna O.* who baptised as the 'talking cure' the therapeutic attempts of her doctor, Breuer. From the point of view of the understanding of the individual, what is the status of this quite particular use of language? For psychoanalysts seduced by the Heideggerian incantation, it is understood that speech is recognized as *being* rather than structure. It is true that one would not find an explicit ontology of language in an author like J. Lacan, for example, but there is the obvious temptation in the form of a kind of 'pangloissism' [universal language] according to which the psyche as a whole is a language, and neurosis is a solecism. In this perspective, the "function of Language is not to inform but to evoke" (Lacan 1968, p. 63). Nothing could more completely close off psychoanalysis' path towards a conceptual understanding than this view.

Such an idea of psychoanalysis tends to enclose the individual in on himself, with the discourse of the patient constituting a necessary frustration, independent even of the silence of the analyst. For, says Lacan, in analysis the subject "finds again the fundamental alienation which made him construct it *like another one*, and which has always destined it to be stripped from him *by another*" (Lacan 1968, p. 11). And the same psychoanalyst protests against a definition of psychotherapy as an adaption of the individual to the social environment: The psychoanalyst should aim at rendering his patient master of his language, not as an instrument of communication, but as the actual substance in the process of his own autonomy. If this is really the meaning and the essential purport of the clinical situation in psychoanalysis, then it would never be anything but magic, and could not give anything of value to a rational science of the human being. But we may think that this quasi-mythical interpretation of speech is by no means the only one offered us.

We must return to the informational idea of language, and instead of opposing to it the image of a purely incantatory, evocative language, it would be good to look for a new, more concrete development of the notion of information starting from the clinical experience of the psychoanalytic couple. On the foundation of a structuralist schema, a dynamic theory of the individual usages of speech, and of its deviant usages is possible: scattered materials already exist for this enterprise, although they scarcely concern traditional pathology alone; structural linguistic conceptions and cybernetic ideas have barely begun to be used to examine the problems of speech. The analytic situation calls for more general research: by somehow making a psychopathology of daily life symmetrical with a synthetic theory of linguistic individuation, which extends at once psychology, social psychology, and the science of language. One can have an inkling of this discipline if it is seen as a theory of redundancy, for it is essentially in this super-addition to the strict informational content that individuation can appear. At all levels of linguistic structuration, an analogous phenomenon shows up: the constitution of a system of 'free' variants which confers on the usage of a phonological schema its stamp, its accent, and on language usage, its style. Such a discipline should tend to clarify the still obscure relations between the verbalized and the unformulated. Clinical psychoanalysis poses the problem which the conjunction of the linguist, the psychologist, and, no doubt, the cybernetician can elucidate. The recent progress of linguistics will thus find itself quite naturally placed in a privileged position, dominating several important sectors of the epistemological program whose traits we have seen become gradually firmer and firmer. The modern conception of language becomes at once a model and an instrument for a concrete analysis of structures and their effective introduction into a situation such as we have characterized.

7.7. There remains a third major theme in the methodology of psychoanalysis; this is that of the relations between psychoanalysis and sociology. If psychoanalysis was presented originally as an attempt to introduce the individual into the domain of science, it quite early encountered the problem of the social being. From our own point of view the most interesting aspect of this encounter appears to be the kind of osmosis that obtains between psychoanalysis and anthropology. It is not the only such aspect, but we consider it representative of the promises and dangers offered by this recent phenomenon of scientific culture.

The elaboration of the clinical situation leads to a more or less valid, dynamic integration of the structure of the ego; this structure of the ego is then taken as a cultural paradigm around which social institutions and actions revolve. From a method for understanding the individual, psychoanalysis becomes a method for the exploration of social life. Certainly this epistemological process is realized in very different fashions by Freud himself, by Jung, by Ruth Benedict, by Kardiner. I shall not attempt to analyze these different aspects. In general, this process interests me to the extent that it claims, in its realization, a sort of Copernican reversal in the sciences of man. The traditional positivist schema, insofar as it admits any knowledge of the individual, in fact assumes implicitly that this knowledge must begin with a science of the general. Psychoanalytic anthropology proposes, on the contrary, an explication of social structures through the clinical understanding of the individual. But this individual is shorn of his essential characteristic of deviant: he is viewed, on the contrary, as a paradigm . . . More precisely, the anthropologist attempts to infer from a single *case* a typical structure, which would be the still empty form of the individual. This is the 'basic personality' of Kardiner. Deviation is then effected in relation to this formal framework; and the pathology appears when the individual contents of the personality cannot be introduced into the mold. As for society, it is no longer described as a set of conditions which determine this structure of the ego, and as a set of products which this structure of the ego generates. In a recent text (Kardiner 1959) the author develops the consequences of this revolution in an original direction. A society, he says, in substance, not being an organism, does not have its own 'homeostats'. From what then derives the stability of social apparatus? Kardiner responds, the 'human unit', that is the individual structured according to the basic personality. This is probably an unsustainable idea in its literal interpretation, but it is seductive and fertile if interpreted more broadly. If in effect one envisages societies as 'great technical units' (in the sense of §§4.20 and 6.8.), it is essentially at the level of individual reality that the informational processes are carried out on which the mechanism of homeostasis depend. In order to understand them, one must consider the technical unit as a whole, and not hypostasize a structure of the ego.

Such a hypertrophy of the psychoanalytic point of view seems open to the following two arguments.

(1) Does not the establishment of individual clinical types, defined as representative in a society, depend very narrowly on the cultural traits of the group to which the enquirer belongs? If so, it must be asked whether without extreme precautions, the clinical method can be utilized as an instrument of intercultural exploration. The present state of knowledge certainly does

not allow us to avoid this essential difficulty. This objection is related to the view of C. Lefort, when he says that the social milieu as a whole reflects a certain configuration of personality, and that the individual and society are in a relation of reciprocal expression and symbolization (Lefort 1951).

(2) The set of social phenomena is apparently difficult to reduce to facts concerning the structure of the individual. Psychoanalytic anthropology returns by a circuitous route to a macro-psychology, ignoring in the end the differences between various levels of structure and claiming improperly to substitute itself for a sociology. Thus, I believe that clinical understanding cannot be the model and the unique source for the science of human facts. Quite to the contrary, it is situated in the epistemological program at a high level of elaboration, at the point of articulation of knowledge and practice. Science can validly aim at the individual only after a very long detour.

The methodological contribution of psychoanalysis to an understanding of the individual thus should not be presented as a total subversion of the scientific ideal. If it contributes effectively to triggering a revision of science, this is no doubt to the extent that the objectivization of the clinical situation calls forth a softening of the models employed in the other disciplines, and puts in perspective, within a practice, the notion of structure.

DIACHRONIC AND SYNCHRONIC: PERSONALITIES AS INFORMATIONAL SYSTEMS

7.8. The same movement is manifested in the attempts of personality theorists to define the frameworks of individuality. And the concurrence between two tendencies is all the more lively as they represent two modes of approaching the identical problem, that of joining structures and the individual. This is a problem which, in both cases, is approached within the perspective of an applied science, it being true that the understanding of the individual makes rational sense only within a practice. But, while psychoanalysis insists on individualization as a temporal structure, whose nodes are events, personality theory insists on individuation as a synchronic structure, whose nodes are the interactions of functions. The opposition between the two points of view cannot fail to recall the opposition between the two linguistics, the diachronic and the synchronic, and a reflection on this analogy should clarify considerably the psychologists' disagreements. The individualization of the being as personality at bottom poses the same scientific problem as that of the concrete constitution of a language. The latter is a system of elements of expression which must be conceived and described by means of informational models, and not simply by means of 'energetic' ones. In an integral linguistics the two levels must be envisaged. The same is true for personality. Psychoanalysis, although it brings to light the informational aspect of the psychic fact, and proposes to treat psychological events as *signs*, remains nonetheless incapable of coordinating this specific structure and the energetic structure of traditional science. Ambiguous, it continues to describe the individual psyche as a system of forces. From this derives the tiresome animism with which most of its schemes are clothed. The fundamental progress to be realized is an operationally valid distinction between an energetic 'infrastructure' and an informational 'superstructure'.

Personality theory encounters the same difficulties, but by reversing the order of preferences, it attempts to treat the components of personality as forces, while describing systems which have value only when interpreted as arrangements of significant elements, of oppositional and relational entities. Now the transposition of linguistic structuralism into psychology presupposes the provisionally radical distinction between diachronic and synchronic descriptions: it seems to us that this is the price of development of a scientific personality theory, capable of assimilating the conquests of psychoanalysis without reluctance.

7.9. The current situation remains very obscure, and two major temptations beckon personality psychology as the concrete science of the individual. The temptation to 'classify' leads the personality theory toward a definition of types in themselves, entities taken in the raw state from experience and vulgar practice. So simplistic an aim satisfies the spontaneous pseudo-platonism that dominates the beginnings of scientific thinking; it obviously can only lead to a mythology, and the popularizations of psychological notions risks giving it credence it could not have with psychologists.

The 'combinatory' temptation is more subtle. It tends to substitute for the elaboration of a real coherent system the pure and simple juxtaposition of traits which are combined systematically and independently. It might be believed that our connecting personality theory and structural linguistics favors enterprises of this sort. But linguistic structures are by no means the result of arbitrary combinations. They assume, as we have seen, the recognition of significant dimensions, and the determination of substructures. Just as clearly as in a system of language, the constitutive elements in the system of characteristics must have relative and oppositional values. The character 'traits' cannot be conceived as an absolute and isolated determination, but must take its meaning by its relation to other traits. In such and such vocalic system, 'rounding' will be a determinant dimension of phonemes, while in another it will not occur, or occur only as a redundant bound variation, accompanying other traits. Similarly, in personality theory the trait of 'emotivity', supposedly operationally defined, will have a different value according to the constellation it enters, and it will be able to play a role either of 'free' significant modification or of a bound variant, like phonological traits. The most circumspect psychologists recognize such a relativity of components of personality but they seem to conceive it only in the 'energetic' style of the relations of forces: such a trait dynamically dominates another. I believe, on the contrary, that we must look to linguistic relations, and introduce the notion of the *relevance* of a variation rather than dominance, thus decisively separating the dynamic from the informational perspective. The dynamic would reveal another order of considerations, it would develop at another stage of the model, which, if desired, would be that of infrastructures, of physiological determinations and relations with the environment. But let my hypothesis be clearly understood: it is by no means a question of dividing the individual into two parts, one of which would never be anything but an epiphenomenon. The distinction which I propose does not depend on an ontology but on a *phenomenotechnique*. A theory of machines could divide its examination of the same technical object into an energetic study and a 'cybernetic' study without assuming an extravagant dualism of the 'being-of-the-machine'. A monistic philosophy is by no means stumped by the distinction of different functional levels. Parallelism, occasionalism, mechanistic materialism are as much metaphysical paralogisms presented as pseudo-solutions to this authentic problem: how to understand and exploit the relations between the two technological levels in the science of the human being.

7.10. If this is so, the Pavlovian typology, which is contrasted with most personality theories by the stress it lays on the equilibrium between the organism and its environment, no longer constitutes in itself a valid solution. On the plane of the dynamic of the individual psyche it seems easily superior to the morpho-psychological doctrines of the autonomous personality, which "monstrously associate what, in the human being, is most subsumable under hereditary determinations, most fixed, most immutable, with what is the creation of the environment, incessant adaption, consciousness and freedom" (Le Guillant and Angelergues 1954). But as soon as the personality is treated as an informational structure, the dynamic of Pavlovian types can figure only as an infrastructure. In order to speak Pavlov's own language, the theory of nervous types remains at the level of the first signal system; the second system of signals is, technologically, of another order; with this system the field of a strictly informational technique is opened up, and it is on this level of informational structures that the second stage of the theory of personality must be built. Thus the Pavlovian doctrine, far from being able to present the definitive state of a Marxist psychology of personality – *definitive* and *Marxist* being in any case a contradiction in terms – must be considered only as a first stage, quite valid as a reaction against the immobility of the 'idealist' personality theories, but utterly inadequate and 'mechanistic' in the present context.

An integrated personality science would thus develop *cybernetic* models of the individual, in which both 'energetic' relations and informational relations would appear together. It would take account of an 'internal' structure, comparable to those Saussurian linguistics has brought to light in language, and at the same time afford a dialectic of events and the environment. Clinical description and structural theory would thus converge toward an applied science, a scientific practice alone capable of providing, in a limited but precise sense, a determination of the individual by concepts.

PRACTICE AS ART AND THE INDIVIDUAL

7.11. All practice is exercised in contact with the individual. I noted previously (§7.2.) that the very development of the natural sciences as applied knowledge introduced this contact explicitly, under distinct forms and at various levels. A physics which becomes capable of predicting and modifying the evolution of determined complexes, a chemistry which creates new elements by dominating a set of subtle conditions: each of these attains objective forms of the individual. The transition from the notion of *system* – that is, a schema aiming at a universal and homogeneous description – to the notion of *model* – that is, the schema of a relatively autonomous complex, in which strategic factors are distinguished – constitutes an essential methodological aspect of this evolution. Of course, this split distillation of the concept fulfilled by the construction of successive models leaves a residue, which is the object of a practice as art. Admittedly, in certain domains, the distillation is so elaborate that this residue vanishes and the practice is entirely conceptualized. The science and technology of raw material more and more perfectly attain this radical reduction of the individual. And it is quite true that at this level the movement of 'normalization' and of mass production tends to efface totally every deviant element, to achieve perfectly interchangeable products, to suppress absolutely in the isolated technical object every trace of the individual. But its annihilation remains a dialectic, for individuality reappears at the level of the man-machine technical complex and it is precisely under this new form that the sciences of man can first of all come to grips with individuality.

The notion of individual seems in fact to have disappeared in the natural sciences with the overcoming of Aristotelianism. Galilean-Newtonian mechanics, classical macroscopic physics, nineteenth century chemistry, all appear to treat of essences, without ever being concerned with the level of individual existences. The *accident*, the event, is apparently relegated outside the domain of science; it concerns the practitioner. But to the extent that the shape of an applied science is realized, the problem of individual realities is reintroduced into the very center of the scientific territory. And it is of course really only at this point that the problem begins to be posed in conceptual terms, in the language of a real understanding, and as a problem susceptible of being effectively resolved.

Let me mention only the general lines of this conceptualization of the individual, providing with this comparison only the beginning of its study in the human sciences. Nineteenth century physics, in its parallel development of phenomenological thermodynamic and statistical mechanics, so to speak, introduced a negative instance of the individual. The individualization of the particle, which is the *empty form* of the individual, and the construction of the statistical model rest on a refusal to give meaning to the individual at this level. The 'metascientific' conclusion, which should be drawn from this microphysics, is that the individual can have conceptual meaning only at the level of macrostructures, as a node of relations. From another point of view, this amounts to saying that *events* occur only on the macroscopic canvas: at the level of particles, everything is an event, nothing is an event; the opposition between event and structure can only appear later. In the perspective adopted by microphysics, it is thus necessary to renounce the description of 'events' like the passage of such and such a photon through such and such a hole in a screen. This at least is the situation in the traditional microphysics during the thirties, and as such was the source of the philosophical paralogisms of indeterminism. For the notion of determinism is conceived for a universe peopled by events and individuals, for which it constitutes the negative foundation, on which individuals and events appear as 'deviants'. The microphysical universe of Heisenberg is too impoverished to outline the opposition of the determined and the contingent, except by an extrinsic reference to the macrophysical universe.[2] Neither space nor time when structured according to our behavior is a suitable framework for the transformations of configurations of particles. The notion of individuality can only be smuggled into it, and cannot find a place there as a concept.

But if modern physics calls upon us to abandon all mythical usage of the individual, in another way it offers us a glimpse of a novel and positive representation of it. To the extent that investigations of physicists and chemists become more precise and more wide-ranging, nodes of convergence appear, more or less complex, more or less consistent, which outline relatively delimited sets. In a very general sense, we are dealing with *technical objects*, not only with completely constructed artifices, but with complexes for whose definition human activity plays such a role, and whose significance in terms of human acts is so determinative that it would be vain to attempt to maintain for their purposes a rigid distinction between nature and artifice. It is precisely this overthrow of the relation between two notions which profoundly modifies the position of the problem of the individual.

In nature, which is traditionally opposed to artifice, and considered as an object of lived experience – *natura materialiter spectata* – everything is, in the final analysis, individual. The natural object becomes the symbol of the inexhaustible, the indescribable. The artificial object, as opposed to the natural, is on the contrary viewed as interchangeable: it participates directly in the nature of the concept, it is schema, reason, and tool. In this sphere, of course, another figure of the individual appears, with the work of art, just as in the sphere of natural objects the point of view of the inexhaustible comes face to face with the system of abstract laws of *natura formaliter spectata*. For both of the two domains, however, it is in the perspective of the experienced that the individual appears, and in consequence, is beyond the reach of scientific determination.

The technical object, in the broad sense that we intend, transcends this opposition between art and nature. Of course, its constitution has been made possible only by this radical split which permitted the original exorcism of the phantoms of magic. But a new phase of science has opened, which makes the separation appear provisional. In this phenomenology of the technical object, a new type of individual is sketched, a conceptualizable type. This new type is, in its roughest form, the individualization of the machine. The simple tool is individual only insofar as it is attached to the lived experience of the user: *the* knife is not an individual, but *my* knife is, evoking my memories. But the machine is already individual, to the extent that it is autonomous and composed of coordinated elements, whose variants can have a functional meaning. A machine is connected, much more closely than a simple tool, and *a fortiori* more closely than a 'natural' object, to the conditions of human existence. Thus it participates explicitly in a history, although in an *extrinsic* way: the human individual participates in a history intrinsically, and that is why his conceptual determination is so difficult and perhaps so limited. Nevertheless, it is the same line of thought which, in the human domain, gives form and content to the idea of the individual.

7.12. There is certainly no question of preserving the experienced tonality of the individual in the concept. The essential epistemological obstacle here is just this refusal to renounce confused ideas born immediately of lived experience, while claiming at the same time to benefit from the clarity, the distinction, the effectiveness of the concept. This appears to be the case in Gurvitch's sociological notions of 'total social fact' and of 'global society'. It is intended to preserve within conceptual thought the images reflecting experience, lived in a purely speculative manner; thus the simple metaphor takes the place of explanation, thus the 'volcanism' of the social fact is introduced into scientific description.[3]

Applied science avoids this obstacle by progressively constituting an articulated concept of the individual at different levels of the object it aims at. This [individual] is a relativized notion, defined in terms of such a structuring and its degrees of freedom. The individual corresponds to the determination of 'free' variants in a system of expression, and generally, to a *redundance*, irreducible in terms of a certain code, of the object considered as 'message'. The progress of scientific knowledge naturally consists in discovering the increasingly fine decoding grids, so that the redundant elements, which are necessarily not the same for all, are diminished in number, without ever being eliminated.

Art, in all its forms, is that phase of practice which concerns unreduced redundance. The physician, the psychologist, the economist as well as the engineer extend conceptual knowledge by an art which treats the individual as an immediate figure of experienced objectivity.[4] The individual, in this sense, marks the limits of science, but only in appearance, for science envisaged as a moment in the total practice coordinates quite naturally and necessarily with art, if it were up to art to interpret the concrete experienced in a system of explanation, as happens, we shall see, for history: but let us leave that singular case for later. The archaic conditions of science can be characterized from this point of view as those whose articulation with practice as art remains inadequate. This was the case for Aristotelian physics and for chemistry before Lavoisier. Through this phase, practice, which can well be rather effective and subtle, remained detached from conceptual understanding; and, linked most often to a mythical 'knowledge', it frequently constituted an obstacle to the development of science itself. Thus a decisive moment in the history of science occurs when practice as art ceases to be this obstacle and becomes a powerful instrument of scientific progress. It seems that for the sciences of man, we are only now beginning to perceive the beginnings of this new age.

INDIVIDUAL AND ALIENATION

7.13. It is thus necessary to renounce the inconsistent dream of a science which would enable us to reach the individual, and in particular the human individual, in the same way that he is given to us in experience – while at the same time retaining rigor, precision, and effectiveness. Science is oriented toward the construction of a *concept* of the individual, and this construction makes sense only in the perspective of applied knowledge.

If, by means of these epistemological considerations, we finally examine the general philosophical meaning of the notion of the individual, we encounter the idea of alienation in perhaps a new light. All the great metaphysics are to some degree attempts to describe and to suppress the alienation of man, considered as stemming from the individual character attached to all experience. The object as well as the subject of all experience are, as such, individuals. Alienation consists then not so much in a loss of self in the other, in a perception of the self as necessarily other, as in the paradox of a felt autonomy, allied to a conceived dependence which characterizes individual existence. Thus every individual, as such, is alienated, and all human life is the development and the metamorphosis of an alienation. But this alienation is realized on different levels. The religious philosophies assign a privileged position to ethical and affective existence; orthodox Marxism gives pride of place to the social level, and professes that all alienation is reducible to the economic and political alienation of the member of an exploited class. But if it is quite true that the suppression of social forms of alienation is the preliminary and primordial task, it does not follow that the individual thus liberated must appear at the same time delivered from all alienation. In reality, the illusion denounced here is expressed, in the practice of current socialist regimes, by a tendency to radically suppress individualization; if it were possible, this suppression would in fact resolve the problem of alienated existence, which the suppression of socio-economic alienations is not sufficient to surmount. This is the totalitarian route. But if in fact all existence is necessarily individual, the reduction of individuation can only be apparent. A real communism, without attacking this unrealizable, negative and vain task, would undo only the social alienations; the free individual would remain, on another level, no less alienated in his 'personal' relations with others and with nature, an alienation which no longer expresses itself in the mode of class struggle, but in the mode of passion, pain, and uncertainty.

If, then, the pantheist promise of the dissolution of the individual in the totality of being is not satisfying, must we admit that man has a fundamental and irremediable impotence? A consistently rationalist philosophy allows us another description of the human condition. It shows us, in science, the development of an exemplary partial solution to alienation, by means of the coordination of abstract formal thought and practice. It is practice as art which achieves the resolution, in each domain, of the speculatively insoluble problem of individual alienation. But there are, of course, other sectors of 'art' besides that of technique, and we have already mentioned that of esthetic creation (§7.12., p. 166). In every case, the fact experienced as individual, in which we are involved, is objectified as a 'message'[5] which partially escapes us, because its redundance cannot be deciphered by such a conceptual code. 'Art' plays on this redundance; its sureness is not doubt illusory, its necessity arbitrary; but it delivers us from dissatisfaction and uncertainty. It thus represents, in all free culture – where authority is exercised only with a view to freeing men from collective alienation – the moment of individuality.

Returning now to the epistemological point of view, we must conclude that the notion of individual, always going back to the experienced, is not, strictly speaking, a concept of science; it is the sign of our alienated condition, and the theme of *praxis* rather than of an isolated understanding. A philosophy of understanding that elevates into essences what are only experiences lived in the mode of the individual treats speculatively what depends on *praxis*, and leaves indefinitely unresolved the problem of the individual.

We have seen however how the science of man, by affirming its character as applied knowledge, is directed toward a specific conceptualization of its object which makes possible and fruitful the junction of knowledge and treatment of the individual.

HISTORY AS A *CLINICAL UNDERTAKING WITHOUT PRACTICE*

7.14. If the notion of the individual can be introduced into science only through the mediation of practice, the paradoxical status of historical knowledge appears in all its gravity. It is perhaps surprising that a whole chapter of the present work has not been devoted to this discipline; the time has come for me to explain myself. History, is in fact defined from my point of view (Granger 1955, Chapter 1) as a *pole* of scientific understanding, but in its realization it cannot constitute the *capstone* of a science of man, indeed, strictly speaking, it cannot be numbered among the human *sciences*. For each of them there is posed the problem, with more or less acuteness, of the historical dimension, without there being a need to consider a history as the goal and end of each one of them. For similar reasons, mathematics cannot appear as the perfect form of the sciences of nature, although their mathematization is essential to the progress of their method, and it is precisely from this perspective of mathematization that I have chosen mainly to consider also the sciences of the human reality. Thus the problem of history is encountered only at the limits of formal thought, because this discipline is offered as the understanding of the individual *par excellence*.

In the sense defined above, history is certainly a clinical understanding. But while psychology, sociology, economics . . . to the extent to which they effectively utilize this method, are necessarily connected with a practice, history on the contrary provides the paradoxical example of a *clinical enterprise without practice*. If it is admitted that the preceding analyses (§§7.3. to 7.10.) are well founded, then such a characterization would suffice to set aside and to justify the refusal to consider history as a science on the same footing as the other disciplines. But perhaps it will not be accepted that history is declared "without practice"; it will be said for example that the practice of history is politics. I do not believe this. History no doubt *accompanies* political *praxis*, by furnishing the materials of a philosophical interpretation of the present, but by no means of a technique for action *drawn from examples of the past*. We will shortly return to the place of history in *praxis*. But to consider history only as 'science', we cannot see that it has ever developed into an applied understanding.

7.15. We may say that history, considered as an autonomous discipline, is not, strictly speaking, one of the *sciences* of man, using this word in its full sense. In history one can speak of laws, of variables, of models only by abusing language, thus confusing this activity with sociology, psychology, economics, . . . If science is defined as the construction of effective models of phenomena (cf. Granger 1959) it is apparent that history escapes us, to the extent it proposes not to elaborate models for a manipulation of realities, but to reconstruct realities themselves, necessarily experienced as individuals. At the other pole of understanding these are the realities that vanish, leaving only models which have become objects: for mathematics the real world is no more than the smile of the Cat in *Alice in Wonderland*, floating in the universe of structures. Symmetrically, for history in its pure state, models disappear in a world of events and persons; at the limit this evocation depends on an art, as all students of history see, though not always acknowledge. Of course, the historian remains constantly in a rather ambiguous situation, alternating, with virtuosity, his use of strictly scientific methods, and the almost always tacit recourse to the evocative imagination of the concrete. He is, then, in turn sociologist, economist, linguist, psychologist, employing more or less explicitly outlined formal models. But he can always rightly claim the originality of his discipline, since his aim is always different from theirs. He wants to reach the individual, but by sight only, and never by touch. He is a *speculative clinician*.

In a certain manner, the French historical school of the *Annales* gives us a very brilliant example of this paradoxical situation. (Although its representatives are no doubt not disposed toward our thesis which denies to history the title of science . . .) Is it not a question, at least for Lucien Febvre, of restoring to history its true object: daily life, the very content of the individual life? Febvre could have borrowed from Hokusai the charming title for his art: 'Ukiyoe', "pictures of passing life". Does not Fernand Braudel's magnificent *The Mediterranean* . . . (Braudel 1972–73) show, both in the distribution of its exposition and in its processes of analysis, that the work of the historian is constantly divided? On the one hand, there is his scientific usage of structured models – a usage that is scientific but irremediably abstract, since it remains cut off by the course of time from any active penetration into the world to which it applies. On the other hand, there is his imaginative evocation *à la* Michelet, which involves the arbitrary determination, the assurance of choice, and the apparent necessity of the fine arts. But there is no other possibility for understanding the content of the human past: the category of the individual can be rigorously applied only to an actual existence, the past individual escapes conceptual understanding, that is, science, if it is true that science can reach the individual only by its insertion in a *praxis*.

HISTORY AND THE PRESENT

7.16. There is thus for history an esthetic temptation, which turns it towards the novel, and a formalist temptation, which turns it toward the social sciences. The historian's talent consists in navigating between these two shoals. Considered as an autonomous and established discipline, history, rather than a science, thus appears to be one of the arts which characterize a culture. From this point of view it is more than a science, in that it already constitutes technical knowledge, but a technique aimed not at modifying our world, but at building a past for us. However careful it is to be 'objective', the very nature of the human past, attached to individual existences, forces history to be an *ideology*. So used, this word should not have any pejorative connotation, provided that the distinction between what is science and what is ideology is clearly drawn. The ideology of the past has been envisaged by historians since the nineteenth century as an *objectivization*, but it has the same function and the same fundamental aim which at other times was found attached to mythological concepts and to chronicles. The objectivization of the individual of the past cannot have the sense of a complete scientific objectivization; it plays the role of an ideological creation, completely symmetrical in many respects with a prediction of our future; for in the domain of the future the same variants are re-encountered: utopia or positive construction.

7.17. If the historian wants to avoid this situation completely, he leaves the field of history altogether. This is the case of a so-called structuralist conception, emphasized by certain historians of philosophy. For these historians, the problem is one of finding in a text no longer the tissue of influences and the traces of a development, but the architectonic of themes, the network of reversals and renewals, of coordinations and subordinations. This involves in short the intention to place oneself in an informational synchronic perspective for which the object is a system of oppositions and of significant relations. This method of analysis, which is introduced all the more remarkably in the movement of renewal in the sciences of man, was developed quite independently of the work of linguists and cyberneticians. But if structural analysis uncovers a mode of objectivization of human products that is new and powerful, does it still truly deserve the title of historical method? I do not believe so. In an exegesis like that of Descartes, by Guéroult, time plays little or no role at all. . . . A strange 'history', this discipline whose greatest concern is to eliminate all descriptions of evolution, all reference to the diachronic. Structural analysis is not a new conception of history, but a method of scientific knowledge, perhaps the first signpost on the road to a *science of human works*. Perfectly defined in its aim, object, and method, such a 'poiematic' would plainly be a science, an explanation by models of a singularly rich ensemble of phenomena, which would by no means be confused with man's past taken as such, but with the complex products of his imagination and his techniques, considered as objective systems of meaning, in the sense of the theory of information and linguistics.

7.18. But let us return to history. Although the human past insofar as it is an individual past essentially escapes scientific understanding, the historical point of view is nevertheless indispensable to the articulation of science and practice. Thus, it is no longer as an autonomous speculative discipline that history is presented, but rather as *an attitude of setting the object in place in the present time*. In this sense, it is the practical synthesis of structural understanding with the experience of the event; it is an art of the constitution of the present, experienced as a moment of our action in a concrete universe, and as such, it is the objectivization of the individual, but not the scientific objectivization of the past. The rationality of the 'methods' on which the historian depends appears here as the rationality of a *technique*, whose finality is not at all absorbing of the event, rendering it irrelevant and inoffensive in a system of formal meanings, but grasping it in its relative contingency and making use of it. History becomes history of the present, that is, a technique of applied analysis: in renouncing the illusory dignity of speculative science, it reveals itself as the final moment of rational conduct. Either history is pure retrospection, and loses itself in an esthetic; or it is an instrument of conjectural and prospective analysis, and it must then be considered not as a science of the human past, but as an indispensable technique of introducing scientific understanding into time. This is a strange dilemma whose severity historians deny, I know: but it is because the myth of history-cum-science has for a century gotten the better of a true idea of its functions and limits.

It is in this sense however, rather than in that of an abstract dialectic of the past and the future, that Marx's saying must be understood and accepted: There is finally only one science of man, history. All the sciences of man end concretely in a historical technique which applies them to the experienced. So that the movement towards the conceptualization of time, the decisive importance for the progress of the sciences we just noted, appears as the preparation, at each level of objectivity, for the really effective rationalization of knowledge.

7.19. This is a rationalization, which is begun over and over again, whose final recourse to history as a technique of temporal insertion obviously marks its limits. The result is that the conceptualization of time in science, in preparing the moment of historical application, is a negation of history. To the extent that a discipline introduces a scientific status for time, it escapes from history, it annexes to science what was the object only of a historical ideology.

I will not take up the analysis of the broad outline of this conceptualization of time, which I have examined at length in another work on the special example of political economy. Let us limit ourselves to noting that the scientific attitude with regard to historicity seems to evolve according to a rather simple scheme. At first, science is naively diachronic. It describes and explains phenomena as temporal development. We have a 'heraclitean', cyclical, diachronic of progress constituting ideological nuances which as such belong to a history, rather than to a science. It would be easy to find traces of these visions of human reality at the beginnings of sociology.

A second moment is that of synchronic description, purely and simply static. Science rids itself of the ideological element inherent in all historical vision; but at the same time it rejects that thing through which the human reality can be made the object of a rational *praxis*. All individuation, and as a consequence all effective application, becomes impossible. The separation between 'scientific' – speculative – knowledge and the art of acting and governing is brutal.

The third moment involves the return to the diachronic, but by the indirect means of a difficult elaboration of the previously developed static concepts. The diachronic determination of phenomena is not reduced to history, because the fundamental scientific aim of explanation by means of models is preserved. The Pygmalion complex, which drives the model builder to require the model's transmutation into experienced existence, is passed. But if this phase has been reached right now in the domain of natural phenomena, where temporality is easily separated from any historical ideology, this is not the case in the human world. Economics is engaged on this path; linguistics is as well. The model constructed by rational thought carries in itself a temporal structure, which is by no means reduced to an eruptive kernel of Bergsonian temporality. For this would amount to making it into a historical being susceptible, at this level, of only a purely verbal understanding, a simple naming of states experienced in the presence of the phenomenon. Of course, on the other hand, this conceptualized and dominated time cannot be understood as an exhaustive definition of experienced duration: this is an issue without scientific significance. Its epistemological value comes, on the contrary, from its being constructed as form indefinitely opposed to a content, and which depends, as I insisted above, on practice. But the dialectization through concepts of time as it is experienced in individual existences permits the substitution, for the uncertain and punctual practice of the primitive *empiricism*, of well-arranged perspectives, hierarchized plans, and the subtlety of a rational *praxis*.

INDIVIDUAL AND FIELD

7.20. Thus the notion of individual that history, as a speculative discipline, offers us cannot serve as the constitutive category for the object of the *science* of man. It remains the motif of esthetic research, and the material of a philosophical interpretation, which it is tempting but vain to substitute for that rigorous and effective understanding through which the phenomena of nature have already been conquered. It is necessary, however, that the human fact preserve, insofar as it is an object of a science, this characteristic of individuality without which, precisely, any introduction of knowledge into practice becomes illusory. It is thus as a function of this requirement that the evolution of the conceptual tool in psychology, sociology, and all the other disciplines directed towards a scientific understanding, should be understood. In this chapter we have tried to outline some indications of this elaboration. It is not possible in the current state of things synthetically to define the categories which would ultimately constitute in a radical – although by no means definitive – fashion the object of science in this domain. One can only foresee the discovery which is being prepared on various sides. It will mark the coming of age of a knowledge of man for which the notion of a *model of the individual* will have a precise and directly applicable meaning. The diagnosis which has been attempted in this book leads to the statement that we have not yet reached this point. Nevertheless we are authorized to give a bold outline of the path upon which science has cautiously embarked.

7.21. The models of the human sciences seem to be necessarily and essentially *cybernetic* models, as opposed to the *energetic* models borrowed from the natural sciences. In the former, the level of information transmissions is superimposed on that of energy exchanges. We have tried to grasp how the introduction of this new canvas foreshadows the conceptualization of the classical oppositions between mind and matter, substratum and superstructure, by moving them from the world of ideology or metaphysics to the world of science. I believe that for models of this nature, individualization begins to appear as an operational notion thanks to the concept of redundance.

But it seems that this individualization cannot be satisfactorily transposed into concepts if one limits oneself to envisaging the models as isolated constructions. The autonomy of the individual object still represents a mythical element in thought about human realities; it is the residue of a specious philosophy of consciousness, which epistemological progress tends to neutralize. For this imaginary intrinsic determination of the individual, science substitutes the opposition of *field* and *point*, the model aiming to define a network of relations whose nodes correspond to the individual. Borrowed from physics, this notion of field could be completely reworked by the specific phenomenology of the human reality. It appears again, in the work of Kurt Lewin, as a setting of energetic relations susceptible of lending itself to an altogether too crudely mechanistic interpretation. The subject finds himself apparently determined as a point of application, and as a center of diffusion, of forces. But it is with difficulty that so simple an image accounts for the empirically verified asymmetry observed between the field as an *environment* determining the action of the subject and the field as *decorative background* constructed and interpreted by the individual. Here again the distinction between energetic and informational relations opens up new perspectives. The subject, in order to become the object of science, must be thought of not only as a point of intersection of lines of force, but also as a center of decision. No doubt the schemata of informational structuring of the field furnished by the theory of communications, and by linguistics, are still quite rough. However, they already permit the placement of gestalt conceptions as well as those of Piaget, in a new perspective. Piaget himself, with his research team, has begun to draw the consequences.[6]

Thus, human facts, experienced as individuals, are transposed into objects of a science, grasped through the field-cybernetic models which are adequate to represent them.

CONCLUSIONS

7.22. The project I formed in beginning this work was no doubt quite ambitious. In taking for a theme the contemporary movement of the sciences of man, I wanted to present under a new aspect the fundamental problem of a scientific philosophy, hypothetically defined as that of the creation of forms and their introduction into a practice. This seemed to me to be the actual character of the problem of the transcendental. Such an enterprise, realized so to speak '*in vivo*' on the living body of science, could certainly not result in the constitution of a synthetic philosophy of knowledge, such as might be expected from the interpretation of a finished science, that is, a dead one. The results of the preceding analyses are thus fragmentary and essentially provisional. If they are valid, this can only be to the extent that they allow us to acquire a better understanding of the apparent paradox constituted by the growing efficacy and exuberant development of formal thought. Through a sometimes fastidiously detailed examination of certain, no doubt perishable, examples of science, I have intended to show *in vivo* how this difficult but fruitful marriage between combinatory and intuitive thought operated.

It is not really the vicissitudes of this *labor*, described as moments in human history, which have been taken as object, for our problem has been one of philosophical diagnosis, and not of history. Moreover, the essential idea which has presided over our research is the concern to bring to light the epistemological 'facts', by comparing the structures built in different domains of scientific thought. From this point of view, our essay mounts a claim, no doubt unjustified with respect to its content, but acceptable at least with respect to its aim, to play a role in this future discipline as a 'poiematic', that is, at the same time a science, a history, and a philosophy of human works. But the truly rational comparative epistemology which would constitute a chapter of this doctrine would no doubt require that its materials be more distinct; in all research, a theory of nascent states and beginning movements is unquestionably more difficult than a static or dynamic treatment of equilibria and current movements. Philosophically speaking, it is, however, the former which fascinate, and seem to uncover the most precious secrets. This is why we have taken for our theme the sciences of man, which are the most uncertain in their approaches, the least convincing in their results.

The difficulty of the enterprise is thus here, despairingly, a direct function of its interest. Nevertheless, if we feel obliged to recognize and accept the limitations imposed on the success of this program, at least we can attempt to outline the positive results of the specific research on which it was based, namely, the present characterization of an understanding of man considered as scientific knowledge.

7.23. A discipline must be capable of being defined as a science, at one and the same time by its aim, its object, and its method. On the last point, the technology of the 'human sciences' can stimulate discussions and disagreements of detail, but it is past doubting that every sufficiently well informed observer recognizes that the same spirit of control and rigor that reign in the natural sciences, is at work here, thwarted, it is true, by incomparable difficulties. However, such a purely technological guarantee cannot suffice for science; we cannot retain this title for a discipline which is methodologically rigorous but which has neither the aim nor the mode of determination of the object which would satisfy the requisite criteria.

As for scientific intentionality, the goal, we have defined it as the construction of *coherent and effective models of the phenomenon*. The limiting cases of mathematics and history have been put aside (§§7.15. and 7.16.). For the other disciplines – psychology, linguistics, economics, and the various specializations of social science – I have tried to verify that this fundamental intention made itself progressively explicit, giving these disciplines their authentically scientific meaning of applied knowledge. The difficulty arises, as in the domain of the natural sciences, not only from our inadequate means of direct action on the phenomenon; it depends essentially on the spontaneous organization of human experience in a system of meanings and values. Man is, first of all, quite naturally, the object of meditation, and it requires a long labor of rational thought for us to acquire the meaning of and need for a real scientific understanding of the human, which is not confused with a philosophy. The limitation of the goal of the understanding to the elaboration of models directly related to a practice continues nevertheless to produce a scandal, as was produced by the substitution of a mathematical physics, extended by an experimental technique, for a symbolic and *profound* philosophical doctrine of the cosmos. Science, as Husserl noted, must in this sense abandon all claim to 'profundity' (Husserl 1965, *in fine*). From this stems the reticences, the indignation, the refusal, which science provokes when it takes man for an object: as if a science of man had to discredit a philosophy of man, and take its place . . . What it makes impossible and empty, in reality, is this ambiguous part of philosophy which is only a mythology, a bastard compromise between metaphysics and science, a glistening dross but without substance, deposited by irrational thought.

7.24. Such an orientation of the understanding of human facts is found to involve a specific determination of these facts considered as *objects* of a science. This is the third criterion of scientific understanding. Therefore I have attempted to sketch the movement of the constitution of the object that manifests itself in these disciplines, and I have encountered the central problem of the relations between structure and the individual, a problem which is no doubt inherent in all scientific objectivization but which only reveals its meaning and takes on its full scope in connection with the human object.

Every scientific object is necessarily constituted by the opposition of a structure and an experienced content, which refers finally to the individual. But in the human fact, individualization cannot be indefinitely neutralized, as it can be in the other disciplines; here science must succeed in grasping, in a certain way, the individual as such. The models toward which it is oriented properly permit this approach. And if they are distinguished from the physicist's models by other than technical details, it is to the extent that they must more and more satisfy this requirement. The progress of knowledge consists here, in the first place, in conceptualizing the opposition between event and structure; the object of the human sciences being then constituted on two levels, as a *structural object* and as a *conjunctural object*, the radical problem is that of their coherent articulation. I believe that it is possible to recognize that this problem is fundamentally without meaning in a speculative perspective, and can only be resolved through applied knowledge.

An epistemology of the sciences of man, if it is not content with a technological examination of methods, but proposes a categorial analysis of the object, thus leads necessarily to the contemplation of a philosophy of practice.

7.25. There is no doubt that scientific knowledge was first of all thought of as speculation, and that it owed its impetus to a negation of practice. Of course, in its truth science has always been bound to the needs and the concrete means of societies, to political organization, to their economies. But the immediately apparent feature of primitive scientific evolution is the ideology of science. In order to escape from the spontaneous alienations which the very nature of the process of perception engenders, scientific understanding originally developed in the direction of an abstract awareness of forms. The prodigious edifice of mathematics remains the masterpiece of the human spirit in its effort to overcome the natural alienation imposed on it by the material conditions of existence. It constitutes a world of the universal and the necessary, as opposed to the singularities and contingencies of individual experience. That the ideological interpretation of this universe coincides, in societies of pre-capitalist and capitalist structure, with a theological and conservative vision of reality in no way alters their value.

But a radical conversion of the conception of the formal becomes necessary when the evolution of the historical content brings forward the dangers of an idealist fixation of science. To the extent that modern societies are dominated by class struggle and the consequences of technological development, the contradictions of the ancient ideology are manifest, the idealist ontology, which was the expression of a victory over the natural alienation of the individual, becomes the instrument of a social alienation. Formal construction in science must now be thought of as the means for action in the world, and not as the trace of a latent reality, opposed to the appearances of the contingent. But this crumbling of an ideology of formal thought by no means implies that formal thought must be abandoned in favor of an illusory *return to things*. On the contrary, the time has come to liberate formal thought by substituting a positive philosophical interpretation for ideology.

In this perspective, phenomenology appears in a certain regard as the prestigious avatar of a moribund ideology. Is not the return to things through a transcendental subjectivity a last attempt at an interpretation of science as the remedy for the natural alienation of the individual? All the themes, the entire problematic of a modern philosophy of science, are brought to light by the genius of Husserl, but are at the same time somehow sterilized by the refusal to truly integrate science in a practice. If Husserl contrasted what he called science with the different 'rational techniques', it was because he desperately wanted to preserve a speculative ideology. Now, we have seen that the very development in both form and content of scientific thought brings clearly to light the contradictions in such an attitude.

Furthermore, a symmetrical error would consist in interpreting this return to things, which must be substituted for the idealist attitude, as a simple recourse to vulgar experience. This is the temptation to which a certain dogmatic Marxism succumbs, through the hyperbolic fear of the sin of idealism. Rejecting all formal thought, this simplistic materialism totally withers the dialectical tree and reduces science to a pure established fact, artificially disguised as a process of movement. On the contrary, it is necessary that the return to things be effected as work, and that formal thought be one of its tools. Applied knowledge corresponds precisely to that state of science in which the formal developments cease to be viewed as revelations of the essence of things. They only apparently belong to the idealistic process of the phantasmagoric appropriation of the world. They are grasped, in their reality, as moments of the complex act of effective appropriation.

7.26. The concept of practice which corresponds to the present forms of existence includes this structural elaboration of the object, the development of which we have shown in science. There is no doubting this last which constitutes today the central motif of social practice, and I have wanted to study the emergence of this motif in the sector of the human reality, which up till now has been dominated by an empirical practice. It would be to sacrifice completely the philosopher's lucidity to ideology if one were to prophesy the total victory of rational thought. Science, no matter how subtle and flexible the approaches that it invents, encounters a radical limit in its conquest of *praxis*. The behavior of the individual overflows scientifically oriented practice; it remains, whatever the degree of rationalization of the social context, the expression of an empirical process of first alienation and then liberation, of which consciousness is the theatre. One might believe that this situation of the individual contradicts and renders sterile the efforts of a science of man. But it does nothing of the kind, if one is willing to admit that man is not only this individual alienated consciousness. He is a scientific object, to the extent that, viewed as an agent within a world, his activity gives a handle on the construction of structured models.

A science of man does not aim at only a part of man; it acquires an understanding of man as a whole, but on a certain level, in a certain perspective. Insofar as it is a *product*, it is itself involved in the reciprocal process of alienation and liberation which occupies consciousness, under the same heading as works of art, or more exactly, *as* a work of art. From this point of view, it would be interesting to apply to scientific thought the category of *style*, to study in this domain the dialectic of an individualization which would somehow clarify the relations between science and practice.

But science is not only a product, it expresses the movement of laborious transcendence, even within the reality of the world to which man belongs, and which it tirelessly opposes to the illusory transcendence of myth. It is when it is applied to itself that this enterprise reveals its most surprising aspects. In the framework of the present work it has been enough for us to follow in broad outline the work of formal creation of a thought that strives to be the instrument of an objectivization of the human reality.

POSTFACE (1982)*

FORM, LOGICS AND REASONS

to Françoise

(1) Is not the possible attraction exerted by a system of thought without empirical content one of the principal dangers to which knowledge of an object is exposed, an object which certainly appears in the sensible world, but which such a uncertain border apparently still separates from the imaginary? We may assume so, if we consider the importance of the almost totally playful, abstract constructions that have been constantly produced in the field of the sciences of man, and the little reality found in so many theories. If, on the other hand, however, one becomes aware of the enormous mass of 'data' almost mechanically recorded by contemporary societies concerning the collective and individual behavior of men, one can ask oneself with good reason whether the relative sterility of this accumulation does not stem from an incapacity of these sciences to constitute concepts that really give a handle to formal thought. The understanding of nature entered the era of continuous progress and discovered the status that became henceforth its own, only after having determined, in each of its fields, the constitutive categories of its objects, frameworks in which the results of observations and experiments could then be gauged and formal entities could serve as themes for a combinatorics, for a calculus. The moment of formal thought in science is decisive. This derives above all from the condition essential to all strictly scientific knowledge which is to take place in and by means of a system of symbols. As Aristotle stressed, a science is, by nature, transmissible; short of thoughtlessly stretching the meaning of the word 'science', this trait must be considered as definitive and as opposing other forms of knowledge incommunicable, or communicable only in terms of imitation and practice.

But the language of a science is not a simple, neutral vehicle. The radical position that consists in denying the possibility of adequately translating scientific understanding from one symbolic system to another can, in this sense, be accepted as a limiting theoretical thesis giving rise to the character that is here essential to a system of expression. But the very movement of science shows that its progress is effected precisely by successive translations into increasingly flexible and powerful languages, translations which

* This Postface has been especially written for the English Translation.

undoubtedly displace concepts and modify them by transposing them. But in a scientific statement – which necessarily belongs to a given state of science – knowledge is distributed both in content and expression. If formal thought plays an essential role in science, even the most empirical, it is above all because it governs its mode of expression. Thus scientific symbolism has a specific function, whose exposure in each science, and comparative examination at different levels and in different fields, is the condition for an analysis of the *formal content* of knowledge. Naturally one need not conclude from this that science is a *game* of writing, and consists in clever and careful variations on some invariant themes furnished by experience, exercises whose interest would depend on the esthetic satisfaction of a particular need of the human spirit. The deployment of a system of formal thought is manifested in science by the construction of symbolic objects, as it happens for art, for magic and for myth. But the structure of these objects in the last named instances remains essentially latent and plays the role of instrument to produce imaginary *lived experiences* under sensible guises. In science the structure of symbolic systems has been made explicit and presented as an abstract schema of the real. A philosophy of science dedicated to interpreting the vicissitudes of language construction must not be taxed with nominalism insofar as it proposes to bring to light the relation between experience and what, for want of a better term, must be called, paradoxically, a formal content.

(2) The progressive creation of this formal content is both shown and concealed under the historical avatars of the sciences, and its significance can appear, as was noted above, only by means of the careful comparison of different moments and different fields. Such an undertaking, pursued to the necessarily excessively limited extent of which I was capable, convinced me of the unity of formal thought. It is not that we must in any way admit the utopian ideal of a unified science and confirm the hope of expressing all objective knowledge by means of a single language whose primitive elements would refer to elementary experiences of a single type. The fleeting attempt at a 'physicalism' could result only in a defeat. It is not toward a reduction of science to a single fundamental kind of object that the unquestionable progress of our knowledge of the world is tending, but rather toward a diversification of the means of formal thought in its approach to lived experience. Its first great successes – and indeed its only really decisive successes – have been obtained undoubtedly in the field of lived experience that we call physics, given to us through the perception of the external world. But

other areas of lived experience have not stopped offering us as-yet-unconquered obstacles to the constitution of a scientific knowledge. The challenge of a science of man lies in the difficulty of forming concepts based on this lived experience that cannot be reduced to physics, and of associating with these concepts *data* susceptible of a formal treatment. It is not simply a matter of searching in the arsenal of ready-made formal systems for some adequate tool to describe the concatenation of phenomena. The effort of imagination and invention of formal thought must be exerted first of all at a deeper, more elementary level, apparently more difficult to penetrate; it should give form to the phenomena themselves, it should ultimately create new concepts, as the sciences of nature did in the eighteenth century.

The first movement of a system of formal thought in the sciences, in view of this construction of concepts, consists in rendering the natural usage of language *metaphorical*. In the sciences of nature such a change of meaning is by now so common that it is not even really felt. When the physicist uses vocabulary referring in the first instance to the lived experience of perception, he does not speak directly of this lived experience: his discourse refers to a system of abstract objects whose structure is evoked and represented, but not precisely outlined, by the concrete relations of perceived events. In the human sciences, on the other hand, it is particularly difficult to separate a naive usage from a metaphorical usage of notions. The object aimed at, on the level of the experienced, is already charged with significations. It is itself an element of a symbolic system and as such susceptible of a discursive knowledge, immediately present and spontaneous. As directly applicable to the practice of life or as the esthetic generator of satisfactions, this knowledge still could not take the place of a science; it cannot be formulated into concepts except in its simplest aspects, nor can it be developed in a controlled and transmissible way by means of a combinatory logic and a deductive calculus. A science of human facts, however modest and commonplace its beginnings, is distinguished from this knowledge precisely by a formal usage of thought. It demands a radical ascetic exercise, a disconcerting renunciation of the immediate significations which nevertheless constitute what in our experience appears as specifically human. It is understandable that some well-meaning people refuse to take the gamble and denounce as illusory the ambition to construct a science that should start out by stripping its object of what distinguishes it from natural objects. It is also understandable that, by a symmetrical and inverse reaction, certain people think that such a science can be nothing other than the direct extension of the sciences of nature and that human facts are by no means essentially distinguished from other natural facts. It has always seemed to me that both attitudes show a lack of confidence in the powers of the creative thought of concepts; for the very beginnings of the sciences of nature have presented difficulties of the same order. Certainly the role and place of a *practical* knowledge *of the individual* will always remain in the human domain the fundamental conditioning of a scientific knowledge whereas a science of nature could be built as relatively independent, if not of techniques, at least of applications to particular events. Nevertheless, within the limits imposed by this condition, we may believe in the development of a true science of human facts. The hypothesis at the source of this book, perhaps about to be read, is that the crux of the difficulty, the place of the most essential obscurities, is the use of a system of formal thought in this field.

(3) Now the richest and most finished system of formal thought is provided by mathematics where it is used, one might say, in all innocence and in the great majority of cases, *pleasure*. It is undoubtedly reasonable to think that the first attempts at a mathematics were comparable to those of a science of nature reduced to the knowledge of figures of space and counting of objects, the preliminary moment of conceptual abstraction having been already reached following a conquest that almost entirely escapes history. But mathematics really begins when these abstract notions themselves have become objects: with the Pythagoreans, with Euclid, mathematics is already clearly a science with a *formal content* whose presentation is by no means abandoned to an experience, but deliberately entrusted to operative rules which prove what can no longer only be shown. It seems to me that the property of formal thought appears in its extreme form as a dual position of systems of operations and of systems of objects, the inseparable reverse and obverse of the same formal 'reality'. Logicians, in mathematizing on mathematics, have asked questions of themselves about the modalities of this duality, and in different ways have faced and resolved the paradoxes and limitations that mark the boundaries imposed on the creativity of formal imagination. But mathematical thought remains nevertheless always free, alive and fruitful as well. It provides abstract schemas that have above all served as instruments for the sciences of nature; it was legitimate to think that it could equally successfully be a reservoir of forms of the human sciences. But despite local and modified successes, the immediate transposition of tested abstract models in the sciences of nature has hardly yet satisfied our attempts nor the hopes expressed with such enthusiasm, and with as much respect as

lucidity, in the middle of the nineteenth century and already in the eighteenth by a Marquis de Condorcet.

Perhaps one of the reasons for this is that mathematical tools, even when their creation has as a point of departure a need, a problem, an idea of empirical knowledge, are inevitably deployed for their own sake. They engender their own obstacles, they proliferate in an operations-objects system that separates them from the primitive given. It happens, it is true, that as they progress the sciences of nature are posed new problems in sufficiently abstract terms that the autonomously created mathematical forms can be recovered as instruments carefully predestined for their formulation: such as Hilbert spaces, tensor analysis, group theory in contemporary physics. But this is by means of a conceptual maturity already acquired by the system of thought of the physical world; the exercise of formal thought is by no means reduced to the transfer of the mathematician's abstract systems into the description of phenomena: it first of all had to succeed in translating and formulating the elements of the empirical into concepts before being able to use the universe of operations and objects offered by the mathematician in a sensible and fruitful way. The difficulties encountered today by the most advanced sciences themselves in the field of quantum mechanics and subatomic physics perhaps stem precisely not so much from a shortage of tools as from an absence of the elaboration of concepts linked to experimental reality to which the tools must be applied.

In the sciences of the human reality this shortage is still masked by the eagerness with which mathematical innovations are periodically received, as if every invention in this field could suddenly shed light and permit us at last to base a description and explanation of phenomena on adequate categories. It ill becomes the philosopher-observer to poke fun at this infatuation, and in this book I have very much – perhaps too much – insisted, on the contrary, on the advantage of the introduction of new tools. But it seems clear to me today that decisive progress cannot be achieved in this field by merely perfecting technologies. In order for this machinery not to slip out of gear, it is essential that it be constantly tested as to the conditions of its utilization. A critique of the requisites of mathematization remains, then, one of the fundamental tasks of formal thought in the sciences of man. It is undoubtedly because of having failed to pay sufficient attention to this discipline that the different sciences of man, those themselves most engaged in the process of formalization, have only feebly responded to the hopes formulated twenty years ago in this book, when formal thought, in other fields like that of molecular biology, seemed to have reached its level of full utilization.

We return then to this idea that the fruitful deployment of formal thought in the human sciences, if it undoubtedly demands an adequate mathematics, requires above all a conceptual preparation of the phenomenal field. Contrary to what could sometimes happen in physics, mathematical concepts do not seem to be able to directly furnish schemas and reference systems for the signifying facts with which science is concerned.

But the difficulties encountered in this field only go to show more especially the complementarity of an algorithmic manipulation and a conceptual determination. *Such a duality characterizes rational thought in general.* I intend to sharpen the meaning of this duality with a few words, in order to place the play of formal thought in the science of human facts in a broader perspective.

(4) Let us call the formalizing aspect 'logical', and call 'rational', in the strict sense, the aspect we have up to now termed 'conceptual'. We shall limit ourselves here to recognizing their complementarity according to two dimensions both of which seem essential to me. One concerns the categories of the object and the coordination of a content and a form; the other concerns procedural categories and the subordination of a tactic to a strategy.

What allows us in fact to distinguish first of all two aspects of rational thought in the broad sense is the ideal of the constitutive formal reduction of its logical aspect. This ideal can be characterized by the threefold aim of vacuity, necessity and closure; their meaning must be indicated briefly.

The desire for emptiness in logical thought appears at its most radical in the reduction of an experience to the position, or the rejection of the formula that expresses it. In a certain way, and paradoxically, logical thought is, at the outset, a homage rendered to empiricism: it constructs the form of knowledge based on the single form of a [positively] established fact [*constat*], at the cost of an indifference directed toward every determinate [positive] verification [*constatation*]. The propositional calculus – which could just as well be called the calculus of '[positively] established facts' [*constats*] constitutes the hard core of all logical thought. Not long ago Ferdinand Gonseth could call it the 'physics of the common object', which has the great merit of underlining its objectifying character. Now, far from presenting itself originally as the free creation of forms of a language, the formulation of logical rules empties experience of its contents only in order better to bring to light the solid shapes of a world. Certainly the movement of the liberation of forms continues afterwards in the direction of a mathematics which constitutes them as imaginary objects by recreating formal *contents* on different planes. But if a strictly logical system of thought exists, it is that which continues to watch over the maintenance of forms of the object in general, the down-to-earth-object, the 'classical' object to which, in the final analysis, one is always referred.

One might say, perhaps, that the very development of the positive sciences and in particular of physics, forces us to think of the notion of 'object' in another way. No doubt so, but then we are dealing with a concept situated very high in the hierarchy of constructions, even if it is presented as the most suited to represent an elementary and fundamental system of physical reality. In order to manipulate it by acts of reasoning, we must nonetheless go back down to the objects of the classical type of thought, to statements that are posed or not, to values of functions (even of random functions) which are this or that. The only thing displaced is the field of application of this set of logical tools which do not operate more directly in the area of experience but at a certain level of symbolism. Likewise one could well build deviant logics of every kind: these are the theories of new formal objects, of specific 'geometries' in whose deployment the play of established facts is finally regulated, however, only by the primitive calculus, and parodying a saying that Politzer applied to psychologies, I would dare say that logics are non-classical in the sense that converted savages are Christians.

The ideal of necessity that orients logical thought shares the same desire to identify, at a profound level, the rules of a symbolic game, and the most general constraints taken as indications of the real. Here the property of the logical is undoubtedly that of transposing experience from the lived constraint into the evidence of exhaustivity of a combinatory logic: a combinatory logic of [positively established] statements void of any content, which shows what is simultaneously posable and what is not, and constitutes the zero degree of a 'geometry' of 'solid' figures of a possible world. These are the tautologies of the propositional calculus and the evidences of the positions and rejections that govern the use of the words 'all' and 'some'. But these primary constraints, derived from the examination of potential or effective configurations are transposable into rules bearing only on assemblages of signs and governing the transition from one to the other.

Logical necessity appears then as the result of a policy imposed on the series of certain symbolic assemblages, by virtue of demands that could, in an extreme case, be completely arbitrary. This is just what the axiomatic presentation of logical forms can lead us to believe, through the possibility of an apparently non-motivated choice of initial assemblages of signs, subject only to grammatical constraints, and of inference rules that allow one to produce new assemblages. This is the illusion, however, of a unilateral formalism. The logic of thought always consists in fact in the use of a duality between this regulation of the series of symbolic operative acts and the system of constraints read in a universe of abstract objects whose basic foundation is constituted by the combinatorics of the posed and the non-posed that outlines the framework of every possible objectivity. And the formal systems that exhibit this logical foundation of rational thought in different styles *make sense* to the extent that they postulate that operations, by means of their combinations, delimit the positions of objects, and that the objects determine as potentialities the operations themselves. To speak of a representation or of an interpretation of a formal system in terms of a universe of objects perhaps runs the risk, by placing logic on the same plane as mathematics, of masking what seems to me to be the former's only profound distinctive trait, namely this complete reciprocity of operation and entity; the object, insofar as it is logical, is nothing more or less than one node of a definite system of operations; the operation, insofar as it is logical, exhibits nothing more or less than one of the degrees of freedom of the relation of an object to other objects of the system. At the most elaborate level of abstraction, the object without qualities is no longer anything more than 'posed', 'non-posed', and we live and make experiments such that correlative operations are those of the classical calculus; however, were we to modify its rules by means of a sort of hyperbolic Cartesian artifice, we would notice that the old calculus reappears at the level of metalanguage as soon as one wants to treat the formulas and symbols of the first as objects . . . Thus the necessity desired by logical thought could not have the arbitrary character of a grammar at the very time when it can be presented under an exclusively syntactical form, for it is the name given to the rigorous duality of the operation and the object which delimits the strict domain of the logical exercise of thought.

This leads us naturally to the third logical wish for closure. By this I understand not this or that global property of completeness defined with precision by logicians to characterize formal systems, but the still undifferentiated general notion of a closing of the system on itself. It seems that thought is recognized as logical in the strictest sense only if it postulates, demands or implies that the application of its operations in conformity with the rules can be repeated indefinitely, and that otherwise and dually the objects so defined do not leave its field, as a consequence of neither an indetermination, nor an incompatibility. The first formulation of this ideal is undoubtedly the requirement of non-contradiction, which consists in refusing the appearance of the object noted as 'p and non-p' or the possibility of two correct deductions, one leading to the formula 'a', the other to the formula 'non-a'. But to the extent that reflections on logical thought have developed and that the duality of operations and objects has appeared more clearly, other, different aspects of this ideal of closure have been formulated and a good part of the work of logicians has consisted in sharpening the means to guarantee them. As we know, only the system of the propositional calculus fully satisfies, without mutilation or artifice, the different forms given to the closure requirement. It thus appears to us as the perfect paradigm against which the degree of a theory's 'logicity' can be evaluated, so to speak. The predicate calculus itself, with its rules for the usage of 'all' and 'some', already suffers from a lack of closure in that no finished decision-making procedure is offered in general to show the provable or refutable nature of its formulas, the consistency or inconsistency of its objects. This is because the predicate calculus already extends beyond *strict logical* thought in that it is the theory not of the object in general but of an object whose content is structured by the distinction between individuals and predicates. All the more when this object is enriched, the ideal of closure becomes inaccessible in all its aspects, as formal thought renounces the extreme poverty which logical thought must maintain.

If we agree to so delimit the logical by a radical characterization of the formal, reason begins to appear to us first of all as a regime of thought which certainly goes beyond the formal. As we have said, strictly logical thought is without content, to the extent that it is applied by nature to *virtual* contents; without losing any of its validity, it can slip completely out of gear, if the context in which it is deployed offers no empirical or even conceptual given that could fill in its gaps and actualize its contents. The rational that I wish to distinguish here from the logical is, on the contrary, never detachable from a content. It is all right to interpret the Kantian distinction between a 'faculty of rules' and a 'faculty of principles' in this sense. The 'rules', which constitute the logical side of thought, govern operations whose system determines only the form of objects on which they have bearing. The axioms from which one starts derive none of their value from a content: they only outline, in advance, the framework of an *a priori découpage* of the operative field; they deserve the name 'principles' only metaphorically since we know how to establish different but strictly equivalent systems as to consequences, and it is even possible to economize on it completely as happens in the present 'combinatory' presentation of logic, which is all rules. On the other hand, in strictly rational reasoning, thought uses authentic principles that distinguish and hierarchize degrees of reality and validity in a colorful, three-dimensional world of which the logical scheme intentionally retained only a monochrome and flat chart. Certainly, reason *applies* logic, but it starts from more or less concrete situations, the grasp of which admits asymmetries and privileges, where the objects of the operations of thought have a content recognized in advance as irreducible, which can make the simple combinatorics of operations impracticable. Also the rational *stricto sensu* series of our thoughts aims not so much at a proof – that is, a chain of reasoning conforming to rules – as at a justification. The ordinary use of the expression 'to give reasons' marks quite clearly this difference from the strictly logical approach. It is good to note first this truism: that one gives the reasons not only of a declarative statement but also of an order, a question, an action (considered implicitly as a sign, in a 'language' only foreshadowed by the interlocutor). To give the reasons for these things is first of all to express the *topics* of the situation; one chooses a certain perspective among all those whose convergent multiplicity makes an individual situation of it and one designs a mode of connection that links it to one of three great types of justification evoked by the words 'premise', 'cause' and 'end'. Only the first refers to a logical determination, but in this case it is not its logical character that lays the foundations of its rationality; it is the choice of a logical justification in such and such a situation that is rational or not. In this sense, reason is always *practical* in that it is an art of applying thought to situations. It could not consist in following rules nor could it reduce itself even to a set of meta-rules insofar as such a word can have a meaning. In a certain way reason dominates logic since, by positing principles provided with meaning in concrete situations, it decides to choose the abstract mode of logical sequence or not, to describe it or exploit it. But in another way, however, logic remains essential to reason, for, whatever the type of justification decided on, the discourse that develops is, so to speak, subject to the second degree, to the formal constraints of logic. Thus to think rationally about a human phenomenon can very well consist in recognizing that the statements that transcribe it are not linked in a logical system; it is still true nonetheless that the series of second-order statements, whose organization constitutes the *theory* of this system, will conform to logical rules. Thus non-contradiction remains an essential constraint to rational thought not because reason would impose its presence in the content of the situations it envisages, but because the discourse it holds on the modes of concentration that it assigns to them is a logical discourse, in a space of consistent justification.

(5) Logic and reason are then dual modalities of rational thought in the broad sense, insofar as the first radically subordinates the object to the operative form, and the second, the operative form to the object. They are

still this in that their correlation is that of a strategy and a tactic. Logic in effect concerns *local cycles* of operations, at least in the most restricted sense that we have proposed. The 'decidability' that characterizes it means that a standard procedure exists, which permits us to establish the validity of a formula, or to refute it, in a finite number of steps. For the logical in the broadened sense, including the first-order predicate calculus, we know only that such a tactic exists *if* the formula is valid, but if the formula is not valid, it may be that it needs an infinite number of steps to demonstrate its invalidity. Nevertheless this ideal, of a universal tactical program even thus weakened, seems to me to be constitutive of what is for us the logical, as Leibniz had seen it. We now know that it is very far from being realized, as he believed, for every kind of calculus and this is why above I proposed such a Draconian restriction of the strictly logical field of thought. In a symbolic system that extends beyond this field, every proof demands that one imagine a strategy that organizes tactical algorithms, that one invent intermediary stages, that one subordinate sub-programs to a view of the whole of the calculus. Mathematics, 'free mathematics' as Cantor said, although exclusively logical in its tactical approaches, goes beyond logic as a strategy of creation. In this sense it is already a work of reason. The style of Euclidean and Archimedean rigor tended to hide this architectonic aspect, so that Hegel took this style at face value when he reproached mathematics for the blind unfolding of its proofs. But this is only apparent, for a careful study of mathematical works always reveals a strategic organization whose manifest or latent heuristic principles characterize a style. They serve as a guide for the localization of difficulties and for the form to give to the solutions. But we would search in vain to formulate the *universal principles* of a mathematical reason: here as everywhere reason commands an indefinitely renewed art of execution.

In less abstract fields, rational strategy is opposed to logical tactics in that the first always calls for a confrontation of symbolism and experience. Empirical scientific knowledge develops abstract models whose exploitation is logico-mathematical. The strictly rational element appears first in the very construction of models but also in the continued control of the leeway available in relation to the experience almost fatally imposed by this exploitation. When concept and scientific theory have only a purely logical status, they run the risk of changing spontaneously into machines running with their gears disengaged. But rational thought, in reintegrating them into less arbitrarily *découpé* totalities is forced to recast them and thus to restart the analyses and construction. Reason dominates logical thought in that it gives a direction to its unfolding, it assigns it a goal, it hierarchizes its effects. We could believe, perhaps, that higher-level algorithms are capable of imitating the architectonic power of reason. It is in fact possible to reduce the *optimization* of a sufficiently well defined process to a calculus. But the algorithm of optimization, or more generally of regulation is then only the instrument of this putting into perspective, which depends on the rational act through which the optimizing variable and the strategic factors taken as independent variables have been chosen. It is an act of thought that is not subject to pre-established rules, but applies principles whose effective content cannot be separated from a situation. Likewise, to want to enunciate principles of reason in general is an illusory undertaking. Such principles, detached from their circumstances of application, would fall to the rank of false logical axioms, leaving us to believe in a unicity, a spareness, a rigidity which are by no means the attributes of strictly rational thought. A rationalist philosophy that would define the real in terms of its adequation to these principles would rapidly become untenable. In the final analysis it would not distinguish between logical rationality and 'rational' rationality, simply superimposing on ordinary logic a super-logic of the same nature. It seems to me that it must be said that the real is neither logical nor rational. The only thing that can be logical is the form of an abstract discourse (but associated with elementary conditions of a perception and manipulation of objects). The only thing that can be rational is an approach of thought that wants in the best way to dominate the relation of symbolic constructions to experience (but it does not depend on immutable and closed principles).

Logic and reason, thus defined, could not be considered as two independent modalities of thought, which could be put into practice one without the other, or of which only one would be the distinguishing mark of humanity. Their articulation, it is true, is susceptible of forms of varied equilibria in which the rational moment and the logical moment can in turn have the largest part. A description and a scientific explanation of human behaviors should cause this duality to appear. We seem already to have an apparatus of concepts suitable for building models of logical activity for different stages of its formation. We barely have satisfying concepts by means of which we could try to think empirically of strictly rational activity. It is, however, the search for such concepts that tacitly or overtly animates the deepest currents of modern philosophy. But under the name of dialectics we are given arbitrary pseudo-logics, or visions of the world. The future belongs perhaps to those who will know how to formulate the modalities of the usage of symbolic thought, and of language in particular, into concepts without getting lost, as we too often do complacently, in the open labyrinth of its forms. In any case, if philosophy could inspire some progress in the positive sciences of man, this would undoubtedly be in preparing the ground for the conceptual expression of the relations between logic and reason, in furnishing the proper instruments to define clearly, in our behavior, the trace of the rational.

GILLES GRANGER

'Cassiopée', 1982

NOTES

PREFACE

[1] The image comes, as we know, from Wittgenstein, and concerns what he calls 'things'. In a sense, the *Tractatus* moreover develops an original structuralist conception of language.

[2] For example, Bourbaki, *Theory of Sets* (1968, Chapter 4, § 1, No. 4). Here is a semi-intuitive and slightly simplified interpretation. Given determinate objects: E_1, E_2, . . . E_i, . . . E_n, which are sets, one assumes as accepted fact a so-called theory of sets or a stronger theory T which includes it.

By means of the operations of this theory one constructs with the E_i's a complex set S whose generic element is called s, and it will constitute the fundamental concept of the structure.

A *relation R* is then posited between the E_i's and the s's suitably independent of the choice of E_i's (Bourbaki says 'transportable' and formally defines this term). This relation is the specific *axiom* of the structure considered, and determines the s's which characterize it. For example, for structures of order, there is a unique E and the s's are part of the Cartesian product $E \times E$, satisfying the conditions of transitivity and reflexivity which the axiom states. For topological structures, the s's are parts of the set of the parts of E, having been determined by the axiom as 'open' sets.

[3] I mean that in mathematics everything must be capable of being explicit and formulated. Obviously we are not dealing here with 'closure' in the various technical senses given it by mathematics.

[4] This is already the language of this book, *Formal Thought and the Sciences of Man*: the opposition is, however, formulated and developed in later writings. (For example, Granger 1965.)

[5] In a monograph on the development of structuralisms, one place would obviously be made for the fine works of comparative mythology of G. Dumézil whose concern to find the same meaningful schemas in various myths partakes perhaps more still of a conception of signifying systems – in the spirit of Guéroult – than of a contemporary analysis of structures – in the manner of Claude Lévi-Strauss.

[6] The qualification 'semantic' for such organizations is obviously very troublesome. It is, nevertheless, made legitimate by the fact that a phonological or 'graphological' system is always the indispensable substratum of a symbolization.

[7] What, for example, Michel Foucault, in his work on *The Order of Things; an Archaeology of the Human Sciences* (1971) is careful not to lead one to believe. Whatever objections might be made to the application of his project, at least its design is clear and sound.

CHAPTER I

[1] Summum crede nefas animam praeferre pudori
Et propter vitam vivendi perdere causas.
(Juvenal, *Sat.* VIII, 83–84)

Count it the greatest of all sins to prefer life to honour,
and to lose, for the sake of living, all that makes life worth having.
(Juvenal 1940, p. 165)

CHAPTER II

[1] The technical terms 'grouping' and 'operation', as used by Piaget, may be defined as follows:

GROUPING – This is a hybrid mathematical model or structure originated by Piaget evolving from the two mathematical structures known as a group and a lattice. In Piaget's theories he uses groupings as basic models for cognitive structures. A grouping is, therefore, more complicated than a group since it incorporates the lattice concept as well as the group concept. (A lattice is formally defined as a partially ordered set L in which each pair of elements has a greatest lower bound and a least upper bound. Unfortunately, this definition is of very limited value, except to someone trained in higher mathematics, since a lattice is an abstract mathematical structure and the definition itself contains several technical terms. Perhaps it will help to say that an example of a lattice is the set of all subsets of some universal set, where the partial ordering is the subset idea; the greatest lower bound of any two subsets, A and B, is $A \cap B$, and the least upper bound is $A \cup B$.)

OPERATION – This is perhaps one of the most basic concepts in the works of Piaget, a concept which he further distinguishes into formal operations and concrete operations. First, by operation in general Piaget means cognitive actions which are organized closely together into a strong structure. For example, the sorting of a set of objects into subsets with common characteristics (color, size, shape, etc.) is one kind of operation. The union or intersection of sets are other operations, and so are the usual operations of ordinary arithmetic. Flavell says "any representational act which is an integral part of an organized network of related acts is an operation" (Flavell 1963, p. 166).

Concrete operations are those performed by a child on what is immediately present or observable. They refer, indeed, to the concrete or the actual.

Formal operations, on the other hand, deal with the possible or potential, not just the real or actual. They are concerned with hypothetical-deductive thought which is propositional in nature. They are operations on operations or, as Piaget sometimes says, "operations to the second power."

[From: *The Origin of the Idea of Chance in Children* (Piaget 1976, pp. 250–251) by permission of W. W. Norton and Company.]

[2] Arithmetic signs are employed here analogically to represent operations on classes. Mathematicians and logicians more commonly use $A \cup B$ to designate union.

[3] The mapping into a set of the set product of two sets, taken in a determined order.

[4] I have used the word operation in quotation marks each time that I reproduce the author's ambiguity.
[5] We are dealing here with marking rather than measuring, for the unit of information is obviously arbitrary. The function effectively chosen – binary logarithm – assumes that this unit is equal to the information carried by one sign in a *binary* language.
[6] Recall that the binary logarithm of a number is the exponent of a power of 2 equal to that number. The choice of this function enables one to compare the quantity of information in any dictionary to that of a binary one, which is the minimum dictionary. The information thus expresses the number of *yesses* and *noes* that a *binary* message carrying the same content should have.

CHAPTER III

[1] Recall that 'pragmatics' designates a study of language, which takes into account its relations with the speaker and his audience. This in contrast to a *semantics*, which considers only the relations between signs and the objects to which they refer, and a *syntax*, which considers only the relations of signs to one another (Morris and Carnap).
[2] In addition to the classical works of Berthelot, consider, for example (Taylor 1949).
[3] We have simplified Lavoisier's equation, into which he also introduced, besides water combined with nitric anhydride, a solvent, and a gaseous release of the anhydride.
[4] Lavoisier contributed to a revision of Guyton's nomenclature, which was presented to the Académie des sciences in 1787.
[5] The theory involved in the preceding example is related to the notion of electrovalence, and was exploited by Alfred Werner, Nobel Prize winner in 1913.
[6] For a more general conception cf. §5.6.
[7] I am not making the naive claim of assigning limits to the prospects for cybernetics. I simply want to note in a more concrete fashion that a syntax, fed into a machine, cannot give back science. But it is not inconceivable that machines organized on *several linguistic levels* could further the scientific process more than the calculus, by directly sketching the structural interpretations of phenomena (cf. MacKay 1956).

CHAPTER IV

[1] Among ethnologists, Lévi-Strauss was one of the first to become aware of this situation.

> . . . we are led to conceive of social structures as entities independent of men's consciousness of them (although they in fact govern men's existence), and thus as different from the image which men form of them as physical reality is different from our sensory perceptions of it and our hypotheses about it (Lévi-Strauss 1963–1976, I, p. 121.)

[2] I have tried to interpret Pascal's wager in this sense, and to confront its probabilistic conception with that of game theory.
[3] That is, the product of gain by its probability.
[4] The principle is called 'minimax', for the adversary attempts to *minimize* with respect to the variable which he controls the function of gain that I am trying to *maximize* in relation to my own variable. Cf. below 5.16 and 6.20.

5 The dialectization of the qualificative will be studied in detail in the next chapter. As you see, it is introduced at the very level of the *découpage* of facts.

6 In English the phonological oppositions of atonic [unstressed] vowels are partially neutralized. Similarly in Chinese one finds oppositions of *tones* for elements without prosodic accent.

7 The mellow/strident opposition corresponds to the predominance or rarity of irregular vibrations, i.e., noise. The initial affricate of the German *zahl* is strident, in opposition to the *mellow* occlusive /*t*/.)

8 Cf. Locke and Booth (1955); Garvin (1956); Panov *et al.* (1958); Delavenay (1959); Mounin (1963; 1964).

9 This model leads to the formulation of 'Lanchester's law' or the law of squares: the force of an army is proportional to the square of the total strength engaged.

10 If $T_S = 1/\mu$, $S_0(t) = e^{-\mu t}$ with a probability density of $S_{(t)} = \mu e^{-\mu t}$.

11 Cf. for example Morse (1958) or Pilé (1955).

In Pilé's study, cited above, it is assumed that the average duration of landing is constant, taken as a security threshold. The only random flow is then that of arrivals, and one can show, for example, that the proportion of late planes equals the utilization parameter a, that the probability of a wait of duration t has a simple form: $(1 - a)\, e^t$. The problems of management envisaged are two: first, the average duration of landing being given, what maximum hourly flow rate can be assured, if the probability of waiting longer than some given time must remain below a certain threshold? Second, a rate of arrival being given, what average duration of landing must one expect in order that the probability of a wait longer than a given time does not go over a certain threshold?

12 The first attempts at a game-theoretical account of learning are the work of Bush and Mosteller (1951) and Estes (1950).

13 The schema of games of strategy will be analyzed from the point of view of axiomatization in 6.20 sq. In the present case, the model is what is known as a Bayesian one; but, *for the rat*, who is ignorant of the probabilities of reward and punishment, that is, the strategy of 'nature', it is a model of another kind. The solution (½; ½) assumes that the same absolute value is attributed to 'gain' and to 'lose'.

14 In the model of Bush and Mosteller a third dimension corresponds to the rat's refusal to react.

15 This technique is known by the picturesque name of 'Monte Carlo method'.

16 It is naturally necessary to choose always behavior that will be statistically favored by the strategy of 'nature'.

CHAPTER V

1 Bear in mind, however, the fine chapters of Bachelard's *Essai sur la connaissance approchée* (1927).

2 For example, those corresponding to a structure of *order*.

3 These new objects still constitute a *ring*, but while the ring of natural integers includes an infinity of elements, this ring has only *n* elements.

4 See Bourbaki, *Theory of Sets* (1968) for a much more elaborate definition of the notions of 'morphism' and 'isomorphism', which have a very general scope.

5 Cf., for example, Apostel's interesting essay, 'Equilibre, logique et théorie des graphs [sic]' (1957).

[6] Cited by Lazarsfeld (1954), Chapter 7.

[7] Here I have borrowed my terminology from Condorcet who anticipated this analysis. Cf. Granger (1956).

[8] However, below we shall examine a circumstance comparable to that of the *Sophist*: Guttman scales.

[9] This evocation is quite relative, since, given that the responses are all *verbal* expressions of ordinary language, one can in principle enumerate them. But it is clear that here a practical infinity and continuity is involved.

[10] The group is considered as a statistically representative sample of an ideal universe.

[11] Intuitively, if, on the contrary, $p_{ij}^* > p_i^x p_j^x$, this would mean that the fact of a *yes* response to one of the two questions makes *more probable* a positive response to the other. Clearly, if one considers the set of all opinions, there is no longer any independence between different responses. The association of positive responses (1) and (2), for example, is much more frequent than the yes–no association. This is what gives meaning to the questionnaire. Independence *within one class* expresses in some way the hypothesis that the variation of opinion around a modal opinion are random and not significant.

[12] One compares the experimental frequencies with the theoretical probabilities of different opinions calculated in the structure.

[13] This boils down to envisaging the sums of the *squares* of the deviations from the means, in order to attach the same significance to deviations above and below the mean value. This is assuredly a reasonable way of looking at things; it is not the only way.

[14] In a general manner, a grade w of an original test is then analyzable as a *linear* combination of fundamental grades, x, y, z ... according to each of the factors of the structure:

$$w = ax + by + cz + \dots$$

The co-efficients a, b, c, characterizing the original test in relation to the pure tests are called '*saturations*'.

[15] For example, the global classification of types of society by Gurvitch.

[16] Plato vigorously criticized this practice of the Stranger in the *Statesman*, 262 d.

[17] In my *Méthodologie économique* (Granger 1955) I examined the metrization of the notion of 'utility' in terms of game theory.

[18] Let us remember that the number (whole, rational or real) constitutes only one specific form of quantity.

[19] Moreover, it is necessary that the unit element of this set of operators play the role of unit element in its combination with the originating objects. In elementary algebra, it is the same set of *numbers* which serves both as objects and operators.

[20] Derivation and integration can be defined as linear applications (cf. Bourbaki 1974, Chapter II, § §2 and 4).

[21] I did attempt such an examination in *Méthodologie économique* (1955, *passim*) and in *La mathématique sociale du marquis de Condorcet* (1956, pp. 65–80 and 100–136).

[22] The outcomes can span a *continuum*. In the initial case they can be a denumerable infinity or even finite in number.

[23] Cf. 6.21 and 6.22 where the problem of the axiomatization of decisions is considered by itself.

[24] The existence of a solution of this type, first demonstrated by von Neumann (1928), then by various mathematicians in a simpler way, rests on the properties of vectorial

space of the set of 'mixed strategies' thus constructed. In the final analysis it thus depends on the fundamentally *linear* character of the structure of the game (cf., for example, Vajda, *The Theory of Games and Linear Programming* (1956).

CHAPTER VI

* ['A Formal Theory of Interaction in Social Groups,' reprinted from *American Sociological Review* 17 (1952), in Simon 1957, pp. 99–114.]

1 There has been much comment and criticism surrounding the word '*cybernetics*', introduced by Norbert Wiener in order to designate the science of automatic regulation. In the years since the appearance of his famous work, the notion seems to have been confirmed and sharpened.

The originality of cybernetics does not stem from its study of automata, but from the fact that it brings out the notion of a regulating circuit which transmits *information*, in contrast to the principal circuit of a machine that transmits *energy*. This is what justifies the attachment of the theory of communications to this discipline.

Thus hereafter I shall label 'cybernetic' a model whose essential component is an informational circuit as opposed to an 'energetic' model.

2 This is Simon's treatment in *Models of Man* (1957: Chapter 9, 'On a Class of Skew Distribution Functions,' reprinted from *Biometrika* **42** (1955)). This title shows that the author is interested not only in Zipf's phenomenon; Simon's model aims, in fact, at several other distributions of the same character: distribution of authors according to the number of their publications in a scientific journal, of cities by their population, of incomes by their size (Pareto curves), etc.

3 $k > 1$ and generally close to 2. For small values of the frequency, the function remains satisfied if $f(2)/f(1)$ is close to 1/3 and $f(1)/[\Sigma_1^\infty f(x)]$ is close to 1/2.

4 In fact attempts to study experimentally the law of costs have been made where the cost is equated with the time necessary for comprehension. (Cf. the bibliography in Mandelbrot 1954.)

5 An analogous difficulty relative to the definition of experimental confirmation of a formal law is raised by Carnap in *Logical Foundations of Probability* (1950); the difficulty was borrowed from Hempel. Suppose the two extensionally equivalent 'laws' are:

"All swans are white", or $(x)\ (Sx \dashrightarrow Wx)$
"No non-white thing is a swan", or $(x)\ ({\sim}Wx \dashrightarrow {\sim}Sx)$.

The experimental fact, 'a black goose', confirms the second law without, however, confirming the first. . . .

6 It is this dogmatism about closure which leads Husserl to distinguish 'mathematical' essences from 'morphological' ones (Husserl). This is a distinction which would be fully justified, if mathematical systems were opposed to all others because of their nomological characteristic of closed structures. Since there is no such thing, it is clear that dialectical openness belongs to the strictly mathematical concept as well as to concepts in the other sciences, and that these latter, as much as the others, are capable of exactitude. The process of structuring is one at different dialectically linked levels.

7 In the semantic presentation of the propositional calculus, these are the tautological matrices.

8 In a different sense from that given to this word in the course of the present work.

[9] $H(x) = x \cdot D\log x + (1-x)D\log(1-x)$, where D is a linear operator. In the case of Wiener and Shannon, $D = -1/\log 2$. In the case of Fisher, D is the operator of the second partial derivative in relation to the estimated parameter.
[10] Bernoulli began with a differential condition. He assumed that the subjective increase dy of a satisfaction is proportional to the quotient dx/x of the objective increase by the objective initial fortune x. Whence the integral $y = k \log x$.)
[11] Marginal satisfaction is what corresponds to the acquisition of one additional unit of a good, taking account of the amount already possessed. On the hypothesis of a perfect divisibility of goods and satisfactions, this marginal satisfaction is the derivative of the function of satisfaction in relation to the quantity of commodities possessed. Because of the shape of the curve of satisfactions, this marginal satisfaction in general decreases as the amount of the quantity possessed increases.
[12] Bayes' paper published posthumously in the *Philosophical Transactions*, London (1763, appearing in 1764). These ideas were developed in a paper by Laplace (1774) and involved estimating the probability that a certain cause might intervene to produce an event, knowing that this event has taken place, and supposing the *a priori* probabilities of the different possible causes to be known. These causes correspond to the 'tactics' of nature.

CHAPTER VII

[1] Cf. also P. Luquet (1957, p. 182): "It is desirable to reduce as much as possible the domain of imponderables which intervene in the experienced course of the psycho-analytic cure . . ."
[2] Of course, the notion of determinism is dialecticized and reappears under a new form in the micro-universe.
[3] For a discussion of the problem see 'Evénement et structure dans les sciences de l'homme', *Cahiers* de l'Institut de Science Economique Appliquée, *Dialogues*, No. 1, 1957.
[4] It would be good to develop this point of view with respect to the fine arts. For the artist, art is obviously a *practice*, not a speculation. Art represents a specific exploitation of the individual redundance, parallel to its pragmatic treatment.
[5] It goes without saying that the word 'message' involves no mythical implications, but simply refers to an informational system.
[6] Cf. the works of the Center for Genetic Epistemology at Geneva, for example: *Logique, langage et théorie de l'information* (Apostel *et al.*, 1957); *La lecture de l'expérience* (Jonckheere *et al.*, 1958); *Logique et perception* (Bruner *et al.*, 1958).

BIBLIOGRAPHY OF WORKS CITED

Apostel, Léo. 1957. 'Equilibre, logique et théorie des graphs [sic].' In *Logique et équilibre*, by L. Apostel, B. Mandelbrot and J. Piaget, pp. 119–170. Paris: Presses universitaires de France.

Apostel, Léo, Mandelbrot, Benoît, and Morf, Albert. 1957. *Logique, langage et théorie de l'information*. Paris: Presses universitaires de France.

Bachelard, Gaston. 1927. *Essai sur la connaissance approchée*. Paris: Vrin.

Bachelard, Gaston. 1938. *La formation de l'esprit scientifique*. Paris: Vrin.

Bachelard, Suzanne. 1957. *La logique de Husserl*. Paris: Presses universitaires de France.

Bayes, Thomas. 1764. 'An Essay toward Solving a Problem in the Doctrine of Chances,' *Philosophical Transactions of the Royal Society* **53** 370–418.

Bénassy, M. 1957. 'Psychanalyse et psychologie,' *Revue française de psychanalyse* **21** 225–243.

Bergson, Henri. 1889. *Essai sur les données immédiates de la conscience*. Paris: Alcan. See also Bergson 1910.

Bergson, Henri. 1907. *L'évolution créatrice*. Paris: Alcan. See also Bergson 1911.

Bergson, Henri. 1910. *Time and Free Will, an Essay on the Immediate Data of Consciousness*. London: Sonnenschein; New York: Macmillan.

Bergson, Henri. 1911. *Creative Evolution*. Trans. A. Mitchell. New York: Holt. Translation of Bergson 1907.

Bernoulli, Daniel. 1730–31. 'Specimen theoriae novae de mensura sortis,' *Actes de l'Académie des Sciences de Saint Pétersburg* **5** 175–192.

Berthelot, Marcellin Pierre Eugène. 1888. *Collection des anciens alchimistes grecs* . . . 4 vols. Paris: Steinheil.

Berzelius, Jons Jakob. 1811. 'Essai sur la nomenclature chimique,' *Journal de physique* **73** 253–286.

Bloomfield, Leonard. 1933. *Language*. New York: Holt.

Bloomfield, Leonard. 1939. 'Linguistic Aspects of Science.' In *International Encyclopedia of Unified Science*, ed. by O. Neurath, R. Carnap, and C. Morris, vol. I, no. 4. Chicago: University of Chicago Press.

Bolinger, Dwight. 1948. 'On Defining the Morpheme,' *Word* **4** 18–23.

Bolinger, Dwight. 1950. 'Rime, Assonance and Morpheme Analysis,' *Word* **6** 117–136.

Borel, Emile. 1921. 'La théorie du jeu et les équations intégrales à noyau symétrique gauche,' *C. R. Académie des sciences de Paris* **173** 1304–1308.

Bourbaki, Nicolas [collective pseudonym of a group of French mathematicians]. 1960. *Théorie des ensembles*. (His *Eléments de mathématique* **17**, **20**, Part I, Book 1). 2nd ed. Paris: Hermann. See also Bourbaki 1968.

Bourbaki, Nicolas. 1968. *Theory of Sets*. (His *Elements of Mathematics* **1**). Paris: Hermann. Translation of Bourbaki 1960.

Bourbaki, Nicolas. 1970. *Algèbre I*, Chapitres 1 à 3. New ed. (His *Eléments de mathématique* **23**). Paris: Hermann. See also Bourbaki 1974.

Bourbaki, Nicolas. 1974. *Algebra I*, Chapters 1–3. (His *Elements of Mathematics* **5**). Paris: Hermann; Reading, Mass.: Addison-Wesley. Translation of Bourbaki 1970.

Braudel, Fernand. 1966. *La Méditerranée et le monde méditerranéen à l'époque de Philippe II* (1949). 2nd rev. and corr. ed. 2 vols. Paris: Colin. (3rd ed. 1976). See also Braudel 1972–73.

Braudel, Fernand. 1972–73. *The Mediterranean and the Mediterranean World in the Age of Philip II*. Trans. Siân Reynolds. 2 vols. London: Collins; New York: Harper and Row. Translation of Braudel 1966.

Brockmeyer, E., Halström H. L. and Jensen, A. 1948. 'The Life and Work of A. K. Erlang,' *Transactions of the Danish National Academy, Technical Sciences*, no. 2.

Bruner, Jerome S., Bresson, François, Morf, Albert and Piaget, Jean. 1958. *Logique et perception*. Paris: Presses universitaires de France.

Bush, Robert R. and Mosteller, Frederick. 1951. 'A Mathematical Model for Simple Learning,' *Psychological Review* **58** 313–323.

Carnap, Rudolf. 1950. *Logical Foundations of Probability*. Chicago: University of Chicago Press.

Carroll, Lewis. 1963. *Alice's Adventures in Wonderland* ... (1865). London: The Nonesuch Press.

Champetier, Georges. 1943. *Les éléments de la chimie*. Paris: A. Michel.

Church, Alonzo. 1956. *Introduction to Mathematical Logic*. Princeton: Princeton University Press.

Condorcet, Marie Jean Antoine ... marquis de. 1785. *Essai sur l'application de l'analyse à la probabilité des décisions rendues à la pluralité des voix*. Paris: Imprimerie royale.

Delavenay, Emile. 1959. *La machine à traduire*. Paris: Presses universitaires de France. See also Delavenay 1960.

Delavenay, Emile. 1960. *An Introduction to Machine Translation*. London: Thames and Hudson. Translation of Delavenay 1959.

Destouches, Jean Louis. 1953. *Méthodologie, notions géométriques*. (*Traité de physique théorique et de physique mathématique* 1). Paris: Gauthier-Villars.

Estes, William K. 1950. 'Toward a Statistical Theory of Learning,' *Psychological Review* **57** 94–107.

Fierz-David, Hans Eduard. 1945. *Die Entwicklungsgeschichte der Chemie*. Basel: Birkhäuser.

Flavell, John H. 1963. *The Developmental Psychology of Jean Piaget*. Princeton, New Jersey: Van Nostrand.

Flood, Merrill M. 1954. 'Game-learning Theory and Some Decision-Making Experiments'. In Thrall *et al.* (1954), pp. 139–158.

Foucault, Michel. 1967. *Les mots et les choses; une archéologie des sciences humaines*. Paris: Gallimard. See also Foucault 1971.

Foucault, Michel. 1971. *The Order of Things: an Archaeology of the Human Sciences*. New York: Pantheon. Translation of Foucault 1967.

Garaudy, Roger. 1956. 'De l'empirisme logique à la sémantique,' *Revue philosophique* **146** 217–235.

Garvin, Paul L. 1956. 'Some Linguistic Problems in Machine Translation.' In *For Roman Jakobson; Essays on the Occasion of His Sixtieth Birthday, 11 October 1956*, compiled by Morris Halle, pp. 180–186. The Hague: Mouton.

Goldstein, Kurt. 1948. *Language and Language Disturbances; Aphasic Symptom Complexes and Their Significance for Medicine and Theory of Language*. New York: Grune and Stratton.

Granger, Gilles-Gaston. 1955. *Méthodologie économique*. Paris: Presses universitaires de France.

Granger, Gilles-Gaston. 1956. *La mathématique sociale du marquis de Condorcet*. Paris: Presses universitaires de France.

Granger, Gilles-Gaston. 1957. 'Logique, langage, communication.' In *Hommage à Gaston Bachelard*, pp. 31–57. Paris: Presses universitaires de France.

Granger, Gilles-Gaston. 1958. 'Le pari de Pascal (argumentation rhétorique et théorie des décisions)'. In *Proceedings of the 12th International Congress of Philosophy*, Venice 1958. 12 vols. Florence: Sansoni, 1958–1961. Vol. 12: *Storia della filosofia moderna e contemporanea*, pp. 181–188.

Granger, Gilles-Gaston. 1959. 'Sur la connaissance philosophique,' *Revue internationale de philosophie* **13** 96–111.

Granger, Gilles-Gaston. 1965. 'Objet, structures et significations,' *Revue internationale de philosophie* **19** 251–290.

Grassmann, Hermann. 1862. *Die Ausdehnungslehre*. 2nd ed. Berlin: T.C.F. Enslin.

Green, Bert F., Jr. 1951. 'A General Solution for the Latent Class Model of Latent Structure Analysis,' *Psychometrika* **16** 151–166.

Guéroult, Martial. 1953. *Descartes selon l'ordre des raisons*. 2 vols. Paris: Aubier.

Gurvitch, George. 1955. 'Le concept de structure sociale,' *Cahiers internationaux de sociologie* **19** 3–44.

Guttman, Louis. 1954. 'The Principal Components of Scalable Attitudes.' In Lazarsfeld (1954), pp. 216–257.

Guyton de Morveau, Louis Bernard. 1782. 'Sur les dénominations chymiques, la nécessité d'en perfectionner le systême, et les règles pour y parvenir,' *Journal de physique* **19** 370–382.

Harris, Zellig Sabbettai. 1951. *Methods in Structural Linguistics*. Chicago: University of Chicago Press.

Hartmann, Heinz. 1959. 'Psychoanalysis as a Scientific Theory.' In Hook (1959), pp. 3–37.

Hegel, Georg Wilhelm Friedrich. 1923. *Wissenschaft der Logik* (1812–1816). Ed. by G. Lasson. 2 vols. (His *Sämtliche Werke* **3, 4**). Leipzig: Meiner. (Later editions 1934, 1948). See also Hegel 1969.

Hegel, Georg Wilhelm Friedrich. 1952. *Phänomenologie des Geistes* (1807). Ed. by J. Hoffmeister. 5th ed. Hamburg: Meiner. See also Hegel 1977.

Hegel, Georg Wilhelm Friedrich. 1969. *Science of Logic*. Trans. A. V. Miller, foreword by J. N. Findlay. London: Allen and Unwin; New York: Humanities Press. Translation of Hegel 1923.

Hegel, Georg Willhelm Friedrich. 1977. *Phenomenology of Spirit*. Trans. A. V. Miller, with analysis of the text and foreword by J. N. Findlay. Oxford: Clarendon Press. Translation of Hegel 1952.

Hitchcock, Frank L. 1941. 'The Distribution of a Product from Several Sources to Numerous Localities,' *J. of Mathematics and Physics* **20** 224–230.

Homans, George Caspar. 1950. *The Human Group*. New York: Harcourt, Brace.

Hook, Sidney (ed.). 1959. *Psychoanalysis, Scientific Method and Philosophy*. New York: New York University Press.

Husserl, Edmund. 1910–11. 'Philosophie als strenge Wissenschaft,' *Logos* **1** 289–341. See also Husserl 1965.

Husserl, Edmund. 1913. *Ideen zu einer reinen Phänomenologie und phänomenologischen Philosophie*. Halle: Niemeyer. See also Husserl 1952.

Husserl, Edmund. 1952. *Ideas; General Introduction to Pure Phenomenology* (1931). Trans. by W. R. Boyce Gibson. London: Allen and Unwin; New York: Macmillan. Translation of Husserl 1913.

Husserl, Edmund. 1965. 'Philosophy as Rigorous Science.' In *Phenomenology and the Crisis of Philosophy*. Trans. by Q. Lauer. pp. 71–147. New York: Harper and Row. Translation of Husserl 1910–11.

Hyppolite, Jean. 1946. *Genèse et structure de la phénoménologie de l'esprit de Hegel*. Paris: Aubier, Editions Montaigne. See also Hyppolite 1974.

Hyppolite, Jean. 1974. *Genesis and Structure of Hegel's Phenomenology of Spirit*. Trans. S. Cherniak and J. Heckman. Evanston: Northwestern University Press. Translation of Hyppolite 1946.

Jakobson, Roman and Halle, Morris. 1956. *Fundamentals of Language*. The Hague: Mouton.

Jakobson, Roman and Halle, Morris. 1975. *Fundamentals of Language*. 2nd rev. ed. (1971). The Hague: Mouton.

Jonckheere, A., Mandelbrot, Benoît and Piaget, Jean. 1958. *La lecture de l'expérience*. Paris: Presses universitaires de France.

Juvenal. 1940. *Juvenal and Persius*. Trans. by G. G. Ramsay. Cambridge, Mass.: Harvard University Press.

Kant, Immanuel. 1907–1909. 'Metaphysische Anfangsgründe der Naturwissenschaft' (1786). In *Kleinere Schriften zur Naturphilosophie*. (His *Sämtliche Werke* 7). Leipzig: Meiner. See also Kant 1970.

Kant, Immanuel. 1970. *Metaphysical Foundations of Natural Science*. Trans. J. Ellington. Indianapolis: Bobbs-Merrill. Translation of Kant 1907–09.

Kardiner, Abram. 1959. 'Social and Cultural Implications of Psychoanalysis.' In Hook (1959), pp. 81–103.

Kemeny, John G. 1959. 'Mathematics without Numbers,' *Daedalus* **88** 577–591.

Kopp, Hermann. 1931. *Geschichte der Chemie* (1843–47). Leipzig: Lorentz.

Kubie, Lawrence S. 1959. 'Psychoanalysis and Scientific Method.' In Hook (1959), pp. 57–77.

Lacan, Jacques. 1956. 'Fonction et champ de la parole et du langage en psychanalyse,' *La psychanalyse* **1**. Reprinted in Lacan's *Ecrits I*. pp. 111–208. Paris: Editions du Seuil, 1966. See also Lacan 1968.

Lacan, Jacques. 1968. *The Language of the Self; the Function of Language in Psychoanalysis*. Trans. with notes and commentary by A. Wilden. Baltimore: Johns Hopkins Press. Translation of Lacan 1956.

Lagache, Daniel. 1969. *L'unité de la psychologie, psychologie expérimentale et psychologie clinique*. 2nd enl. ed. Paris: Presses universitaires de France.

Lavoisier, Antoine Laurent. 1782. 'Mémoire sur la dissolution des metaux dans les acides,' *Mémoires de l'Académie des sciences*, p. 492.

Lazarsfeld, Paul Felix (ed.). 1954. *Mathematical Thinking in the Social Sciences*. Glencoe: Free Press.

Lefort, Claude. 1951. 'Notes critiques sur la méthode de Kardiner,' *Cahiers internationaux de sociologie* **10** 117–127.

LeGuillant, Louis and Angelergues, René. 1954. 'La notion de 'type nerveux',' *Evolution Psychiatrique*, no. 3, pp. 507–538.

Lévi-Strauss, Claude. 1954. 'L'analyse structurale en linguistique et en anthropologie,' *Word* **1** 33–53. This appeared in a revised form as Chapter II (pp. 37–62) of the first volume of *Anthropologie structurale* (Lévi-Strauss 1958–1973). For English translation see Lévi-Strauss (1963–1976), vol. I, Chapter II, pp. 31–54: 'Structural Analysis in Linguistics and in Anthropology.'

Lévi-Strauss, Claude. 1958–1973. *Anthropologie structurale*. 2 vols. Paris: Plon. See also Lévi-Strauss 1963–1976.

Lévi-Strauss, Claude. 1963–1976. *Structural Anthropology*. Trans. by C. Jacobson and B. G. Schoepf. (Vol. 2 trans. by M. Layton). 2 vols. New York: Basic Books. Translation of Lévi-Strauss 1958–1973.

Locke, William Nash and Booth, A. D. (eds.). 1955. *Machine Translation of Languages*. Cambridge: Published jointly by the Technology Press of MIT and Wiley, New York.

Luquet, P. 1957. 'Facteurs de guérison de la cure analytique,' *Revue française de psychanalyse*.

MacKay, D. M. 1956. 'The Epistemological Problem for Automata.' In *Automata Studies*, ed. by Claude E. Shannon and John McCarthy. pp. 235–251. (*Annals of Mathematics Studies* **34**). Princeton: Princeton University Press.

Mandelbrot, Benoît. 1953. 'Contribution à la théorie mathématique des communications.' Thesis, University of Paris.

Mandelbrot, Benoît. 1954. 'Structure formelle des textes et communication: deux études,' *Word* **10** 1–27.

Margouliès, Georges. 1943. *La langue et l'écriture chinoises*. Paris: Payot.

Marshak, Jacob. 1954. 'Towards an Economic Theory of Organization and Information.' In Thrall *et al.* (1954), pp. 187–220.

Martinet, André. 1955. *Economie des changements phonétiques; traité de phonologie diachronique*. Bern: Francke.

Morse, Philip M. and Kimball, George E. 1951. *Methods of Operations Research*. 1st ed. rev. Cambridge: Published jointly by the Technology Press of MIT and Wiley, New York. (First publication was in classified form in 1946.)

Morse, Philip M. 1958. *Queues, Inventories and Maintenance, the Analysis of Operational Systems with Variable Demand and Supply*. New York: Wiley.

Mounin, Georges. 1963. *Les problèmes théoriques de la traduction*. Paris: Gallimard.

Mounin, Georges. 1964. *La machine à traduire; histoire des problèmes linguistiques*. The Hague: Mouton.

Nagel, Ernst. 1959. 'Methodological Issues in Psychoanalytic Theory.' In Hook (1959), pp. 38–56.

Neumann, John von. 1928. 'Zur Theorie der Gesellschaftsspiele,' *Mathematische Annalen* **100** 295–320.

Neumann, John von and Morgenstern, Oskar. 1944. *Theory of Games and Economic Behavior*. Princeton: Princeton University Press. (3rd ed., 1953).

Oettinger, Anthony G. 1955. 'The Design of an Automatic Russian-English Technical Dictionary.' In Locke and Booth (1955), pp. 47–65.

Ombredane, André. 1951. *L'aphasie et l'élaboration de la pensée explicite*. Paris: Presses universitaires de France.

Panov, D., Lapounov, A. and Moukhine, I. 1958. 'La traduction automatique,' *Recherches internationales à la lumière du marxisme*, no. 7, pp. 162–193.

Piaget, Jean. 1949. *Traité de logique; essai de logistique opératoire*. Paris: Colin. (Second edition published as *Essai de logique opératoire*. Paris: Dunod, 1972.)

Piaget, Jean. 1950. *La genèse du nombre chez l'enfant*. By Jean Piaget and A. Szeminska. Neuchâtel: Delachaux and Niestlé. See also Piaget 1952.

Piaget, Jean. 1952. *The Child's Conception of Number*. London: Routledge and Kegan Paul. Translation of Piaget 1950.

Piaget, Jean. 1976. *The Origin of the Idea of Chance in Children*, by Jean Piaget and Bärbel Inhelder. Trans. by L. Leake, Jr., P. Burrell and H. D. Fishbein. New York: Norton.

Pilé, G. 1955. 'Etude des délais des aéronefs à l'atterrissage,' *Bulletin du Sémin. de Recherche opérat*. June.

Plato. 1955. *Gorgias*. Trans. W. C. Helmbold. New York: Liberal Arts Press. Translation of Plato 1979.

Plato. 1979. *Gorgias* (1959). A revised text ... by E. R. Dodds. Oxford: Clarendon Press. See also Plato 1955.

Potter, Ralph K., Kopp, George A. and Green, Harriet C. 1947. *Visible Speech*. New York: Van Nostrand.

Prieto, Luis J. 1954. 'Traits oppositionnels et traits contrastifs,' *Word* **10** 43–59.

Saussure, Ferdinand de. 1974a. *Cours de linguistique générale*. (Bally-Sechehaye-Riedlinger edition, edited by Tullio de Mauro). New ed. Paris: Payot. See also Saussure 1974b.

Saussure, Ferdinand de. 1974b. *Course in General Linguistics*. Ed. by Charles Bally and Albert Sechehaye with Albert Riedlinger. Trans. by Wade Baskin. Rev. ed. London: Fontana. Translation of Saussure 1974a.

Schutzenberger, Marcel Paul. 1951. 'Sur les rapports entre la quantité d'information au sens de Fisher et au sens de Wiener,' *Comptes rendus de l'Académie des sciences* **232** 925–927.

Simon, Herbert Alexander. 1952. 'A Formal Theory of Interaction in Social Groups,' *American Sociological Review* **17**, reprinted in Simon (1957), pp. 99–114.

Simon, Herbert Alexander. 1955a. 'A Behavioral Model of Rational Choice,' *Quarterly Journal of Economics* **69**, reprinted in Simon (1957), pp. 241–260.

Simon, Herbert Alexander. 1955b. 'On a Class of Skew Distribution Functions,' *Biometrika* **42**, reprinted in Simon (1957), pp. 145–164.

Simon, Herbert Alexander. 1957. *Models of Man: Social and Rational; Mathematical Essays on Rational Human Behavior in a Social Setting*. New York: Wiley.

Simondon, Gilbert. 1958. *Du monde d'existence des objets techniques*. Paris: Aubier, Editions Montaigne.

Stouffer, Samuel A. *et al.*, 1949–50. *Studies in Social Psychology in World War II ...* prepared and edited under the auspices of a special committee of the Social Science Research Council. 4 vols. Princeton: Princeton University Press. Vol. I: *The American Soldier: Adjustment during Army Life*, by Samuel A. Stouffer, Edward A. Suchman, Leland C. DeVinney, Shirley A. Star and Robin M. Williams, Jr.

Swadesh, Morris. 1954. Review of *Learning a Foreign Language, A Handbook for Missionaries*, by Eugene A. Nida. *Word* **10** 83–86.

Taylor, Frank Sherwood. 1949. *The Alchemists, Founders of Modern Chemistry*. New York: H. Schuman.

Thrall, Robert M., Coombs, C. H., Davis, R. L. (eds.) 1954. *Decision Processes*. New York: Wiley.

Trân Duc Thao. 1951. *Phénoménologie et matérialisme dialectique*. Paris: Editions Minh-Tân. See also Trân Duc Thao 1983.

Trân Duc Thao. 1983. *Phenomenology and Dialectical Materialism. Boston Studies in the Philosophy of Science*, In press. Dordrecht: Reidel.

Vajda, Stephen. 1956. *The Theory of Games and Linear Programming*. London: Methuen; New York: Wiley.

Vuillemin, Jules. 1954. *L'héritage kantien et la révolution copernicienne*. Paris: Presses universitaires de France.

Vuillemin, Jules. 1955. *Physique et métaphysique kantiennes*. Paris: Presses universitaires de France.

Wittgenstein, Ludwig. 1968. *Philosophical Investigations*. Trans. by G. E. M. Anscombe. 3rd ed. New York: Macmillan.

Wittgenstein, Ludwig. 1974. *Tractatus logico-philosophicus*. Trans. by D. F. Pears and B. F. McGuiness. London: Routledge and Kegan Paul.

Zipf, George Kingsley. 1935. *The Psycho-biology of Language; an Introduction to Dynamic Philology*. Boston: Houghton Mifflin.

INDEX OF NAMES